The History of Physics: A Very Short Introduction

VERY SHORT INTRODUCTIONS are for anyone wanting a stimulating and accessible way into a new subject. They are written by experts, and have been translated into more than 45 different languages.

The series began in 1995, and now covers a wide variety of topics in every discipline. The VSI library currently contains over 550 volumes—a Very Short Introduction to everything from Psychology and Philosophy of Science to American History and Relativity—and continues to grow in every subject area.

Very Short Introductions available now:

ACCOUNTING Christopher Nobes
ADOLESCENCE Peter K. Smith
ADVERTISING Winston Fletcher
AFRICAN AMERICAN RELIGION Eddie S. Glaude Jr
AFRICAN HISTORY John Parker and Richard Rathbone
AFRICAN RELIGIONS Jacob K. Olupona
AGEING Nancy A. Pachana
AGNOSTICISM Robin Le Poidevin
AGRICULTURE Paul Brassley and Richard Soffe
ALEXANDER THE GREAT Hugh Bowden
ALGEBRA Peter M. Higgins
AMERICAN HISTORY Paul S. Boyer
AMERICAN IMMIGRATION David A. Gerber
AMERICAN LEGAL HISTORY G. Edward White
AMERICAN POLITICAL HISTORY Donald Critchlow
AMERICAN POLITICAL PARTIES AND ELECTIONS L. Sandy Maisel
AMERICAN POLITICS Richard M. Valelly
THE AMERICAN PRESIDENCY Charles O. Jones
THE AMERICAN REVOLUTION Robert J. Allison
AMERICAN SLAVERY Heather Andrea Williams
THE AMERICAN WEST Stephen Aron
AMERICAN WOMEN'S HISTORY Susan Ware
ANAESTHESIA Aidan O'Donnell
ANALYTIC PHILOSOPHY Michael Beaney
ANARCHISM Colin Ward
ANCIENT ASSYRIA Karen Radner
ANCIENT EGYPT Ian Shaw
ANCIENT EGYPTIAN ART AND ARCHITECTURE Christina Riggs
ANCIENT GREECE Paul Cartledge
THE ANCIENT NEAR EAST Amanda H. Podany
ANCIENT PHILOSOPHY Julia Annas
ANCIENT WARFARE Harry Sidebottom
ANGELS David Albert Jones
ANGLICANISM Mark Chapman
THE ANGLO-SAXON AGE John Blair
ANIMAL BEHAVIOUR Tristram D. Wyatt
THE ANIMAL KINGDOM Peter Holland
ANIMAL RIGHTS David DeGrazia
THE ANTARCTIC Klaus Dodds
ANTISEMITISM Steven Beller
ANXIETY Daniel Freeman and Jason Freeman
THE APOCRYPHAL GOSPELS Paul Foster
ARCHAEOLOGY Paul Bahn
ARCHITECTURE Andrew Ballantyne
ARISTOCRACY William Doyle
ARISTOTLE Jonathan Barnes
ART HISTORY Dana Arnold
ART THEORY Cynthia Freeland

ASIAN AMERICAN HISTORY Madeline Y. Hsu
ASTROBIOLOGY David C. Catling
ASTROPHYSICS James Binney
ATHEISM Julian Baggini
THE ATMOSPHERE Paul I. Palmer
AUGUSTINE Henry Chadwick
AUSTRALIA Kenneth Morgan
AUTISM Uta Frith
THE AVANT GARDE David Cottington
THE AZTECS Davíd Carrasco
BABYLONIA Trevor Bryce
BACTERIA Sebastian G. B. Amyes
BANKING John Goddard and John O. S. Wilson
BARTHES Jonathan Culler
THE BEATS David Sterritt
BEAUTY Roger Scruton
BEHAVIOURAL ECONOMICS Michelle Baddeley
BESTSELLERS John Sutherland
THE BIBLE John Riches
BIBLICAL ARCHAEOLOGY Eric H. Cline
BIG DATA Dawn E. Holmes
BIOGRAPHY Hermione Lee
BLACK HOLES Katherine Blundell
BLOOD Chris Cooper
THE BLUES Elijah Wald
THE BODYCHRIS SHILLING
THE BOOK OF MORMON Terryl Givens
BORDERS Alexander C. Diener and Joshua Hagen
THE BRAIN Michael O'Shea
BRANDING Robert Jones
THE BRICS Andrew F. Cooper
THE BRITISH CONSTITUTION Martin Loughlin
THE BRITISH EMPIRE Ashley Jackson
BRITISH POLITICS Anthony Wright
BUDDHA Michael Carrithers
BUDDHISM Damien Keown
BUDDHIST ETHICS Damien Keown
BYZANTIUM Peter Sarris
CALVINISM Jon Balserak
CANCER Nicholas James
CAPITALISM James Fulcher
CATHOLICISM Gerald O'Collins
CAUSATION Stephen Mumford and Rani Lill Anjum
THE CELL Terence Allen and Graham Cowling
THE CELTS Barry Cunliffe
CHAOS Leonard Smith
CHEMISTRY Peter Atkins
CHILD PSYCHOLOGY Usha Goswami
CHILDREN'S LITERATURE Kimberley Reynolds
CHINESE LITERATURE Sabina Knight
CHOICE THEORY Michael Allingham
CHRISTIAN ART Beth Williamson
CHRISTIAN ETHICS D. Stephen Long
CHRISTIANITY Linda Woodhead
CIRCADIAN RHYTHMS Russell Foster and Leon Kreitzman
CITIZENSHIP Richard Bellamy
CIVIL ENGINEERING David Muir Wood
CLASSICAL LITERATURE William Allan
CLASSICAL MYTHOLOGY Helen Morales
CLASSICS Mary Beard and John Henderson
CLAUSEWITZ Michael Howard
CLIMATE Mark Maslin
CLIMATE CHANGE Mark Maslin
CLINICAL PSYCHOLOGY Susan Llewelyn and Katie Aafjes-van Doorn
COGNITIVE NEUROSCIENCE Richard Passingham
THE COLD WAR Robert McMahon
COLONIAL AMERICA Alan Taylor
COLONIAL LATIN AMERICAN LITERATURE Rolena Adorno
COMBINATORICS Robin Wilson
COMEDY Matthew Bevis
COMMUNISM Leslie Holmes
COMPLEXITY John H. Holland
THE COMPUTER Darrel Ince
COMPUTER SCIENCE Subrata Dasgupta
CONFUCIANISM Daniel K. Gardner
THE CONQUISTADORS Matthew Restall and Felipe Fernández-Armesto
CONSCIENCE Paul Strohm
CONSCIOUSNESS Susan Blackmore
CONTEMPORARY ART Julian Stallabrass
CONTEMPORARY FICTION Robert Eaglestone

CONTINENTAL PHILOSOPHY Simon Critchley
COPERNICUS Owen Gingerich
CORAL REEFS Charles Sheppard
CORPORATE SOCIAL RESPONSIBILITY Jeremy Moon
CORRUPTION Leslie Holmes
COSMOLOGY Peter Coles
CRIME FICTION Richard Bradford
CRIMINAL JUSTICE Julian V. Roberts
CRITICAL THEORY Stephen Eric Bronner
THE CRUSADES Christopher Tyerman
CRYPTOGRAPHY Fred Piper and Sean Murphy
CRYSTALLOGRAPHY A. M. Glazer
THE CULTURAL REVOLUTION Richard Curt Kraus
DADA AND SURREALISM David Hopkins
DANTE Peter Hainsworth and David Robey
DARWIN Jonathan Howard
THE DEAD SEA SCROLLS Timothy H. Lim
DECOLONIZATION Dane Kennedy
DEMOCRACY Bernard Crick
DEPRESSION Jan Scott and Mary Jane Tacchi
DERRIDA Simon Glendinning
DESCARTES Tom Sorell
DESERTS Nick Middleton
DESIGN John Heskett
DEVELOPMENTAL BIOLOGY Lewis Wolpert
THE DEVIL Darren Oldridge
DIASPORA Kevin Kenny
DICTIONARIES Lynda Mugglestone
DINOSAURS David Norman
DIPLOMACY Joseph M. Siracusa
DOCUMENTARY FILM Patricia Aufderheide
DREAMING J. Allan Hobson
DRUGS Les Iversen
DRUIDS Barry Cunliffe
EARLY MUSIC Thomas Forrest Kelly
THE EARTH Martin Redfern
EARTH SYSTEM SCIENCE Tim Lenton
ECONOMICS Partha Dasgupta
EDUCATION Gary Thomas
EGYPTIAN MYTH Geraldine Pinch
EIGHTEENTH-CENTURY BRITAIN Paul Langford
THE ELEMENTS Philip Ball
EMOTION Dylan Evans
EMPIRE Stephen Howe
ENGELS Terrell Carver
ENGINEERING David Blockley
ENGLISH LITERATURE Jonathan Bate
THE ENLIGHTENMENT John Robertson
ENTREPRENEURSHIP Paul Westhead and Mike Wright
ENVIRONMENTAL ECONOMICS Stephen Smith
ENVIRONMENTAL LAW Elizabeth Fisher
ENVIRONMENTAL POLITICS Andrew Dobson
EPICUREANISM Catherine Wilson
EPIDEMIOLOGY Rodolfo Saracci
ETHICS Simon Blackburn
ETHNOMUSICOLOGY Timothy Rice
THE ETRUSCANS Christopher Smith
EUGENICS Philippa Levine
THE EUROPEAN UNION John Pinder and Simon Usherwood
EUROPEAN UNION LAW Anthony Arnull
EVOLUTION Brian and Deborah Charlesworth
EXISTENTIALISM Thomas Flynn
EXPLORATION Stewart A. Weaver
THE EYE Michael Land
FAIRY TALE Marina Warner
FAMILY LAW Jonathan Herring
FASCISM Kevin Passmore
FASHION Rebecca Arnold
FEMINISM Margaret Walters
FILM Michael Wood
FILM MUSIC Kathryn Kalinak
THE FIRST WORLD WAR Michael Howard
FOLK MUSIC Mark Slobin
FOOD John Krebs
FORENSIC PSYCHOLOGY David Canter
FORENSIC SCIENCE Jim Fraser
FORESTS Jaboury Ghazoul
FOSSILS Keith Thomson

FOUCAULT Gary Gutting
THE FOUNDING FATHERS
R. B. Bernstein
FRACTALS Kenneth Falconer
FREE SPEECH Nigel Warburton
FREE WILL Thomas Pink
FREEMASONRY Andreas Önnerfors
FRENCH LITERATURE John D. Lyons
THE FRENCH REVOLUTION
William Doyle
FREUD Anthony Storr
FUNDAMENTALISM Malise Ruthven
FUNGI Nicholas P. Money
THE FUTURE Jennifer M. Gidley
GALAXIES John Gribbin
GALILEO Stillman Drake
GAME THEORY Ken Binmore
GANDHI Bhikhu Parekh
GENES Jonathan Slack
GENIUS Andrew Robinson
GEOGRAPHY John Matthews and
David Herbert
GEOPOLITICS Klaus Dodds
GERMAN LITERATURE Nicholas Boyle
GERMAN PHILOSOPHY
Andrew Bowie
GLOBAL CATASTROPHES Bill McGuire
GLOBAL ECONOMIC HISTORY
Robert C. Allen
GLOBALIZATION Manfred Steger
GOD John Bowker
GOETHE Ritchie Robertson
THE GOTHIC Nick Groom
GOVERNANCE Mark Bevir
GRAVITY Timothy Clifton
THE GREAT DEPRESSION AND
THE NEW DEAL Eric Rauchway
HABERMAS James Gordon Finlayson
THE HABSBURG EMPIRE
Martyn Rady
HAPPINESS Daniel M. Haybron
THE HARLEM RENAISSANCE
Cheryl A. Wall
THE HEBREW BIBLE AS LITERATURE
Tod Linafelt
HEGEL Peter Singer
HEIDEGGER Michael Inwood
THE HELLENISTIC AGE
Peter Thonemann
HEREDITY John Waller
HERMENEUTICS Jens Zimmermann
HERODOTUS Jennifer T. Roberts
HIEROGLYPHS Penelope Wilson
HINDUISM Kim Knott
HISTORY John H. Arnold
THE HISTORY OF ASTRONOMY
Michael Hoskin
THE HISTORY OF CHEMISTRY
William H. Brock
THE HISTORY OF CINEMA
Geoffrey Nowell-Smith
THE HISTORY OF LIFE
Michael Benton
THE HISTORY OF MATHEMATICS
Jacqueline Stedall
THE HISTORY OF MEDICINE
William Bynum
THE HISTORY OF PHYSICS
J. L. Heilbron
THE HISTORY OF TIME
Leofranc Holford-Strevens
HIV AND AIDS Alan Whiteside
HOBBES Richard Tuck
HOLLYWOOD Peter Decherney
HOME Michael Allen Fox
HORMONES Martin Luck
HUMAN ANATOMY
Leslie Klenerman
HUMAN EVOLUTION Bernard Wood
HUMAN RIGHTS Andrew Clapham
HUMANISM Stephen Law
HUME A. J. Ayer
HUMOUR Noël Carroll
THE ICE AGE Jamie Woodward
IDEOLOGY Michael Freeden
THE IMMUNE SYSTEM
Paul Klenerman
INDIAN CINEMA Ashish Rajadhyaksha
INDIAN PHILOSOPHY Sue Hamilton
THE INDUSTRIAL REVOLUTION
Robert C. Allen
INFECTIOUS DISEASE Marta L. Wayne
and Benjamin M. Bolker
INFINITY Ian Stewart
INFORMATION Luciano Floridi
INNOVATION Mark Dodgson and
David Gann
INTELLIGENCE Ian J. Deary
INTELLECTUAL PROPERTY
Siva Vaidhyanathan

INTERNATIONAL LAW Vaughan Lowe
INTERNATIONAL MIGRATION Khalid Koser
INTERNATIONAL RELATIONS Paul Wilkinson
INTERNATIONAL SECURITY Christopher S. Browning
IRAN Ali M. Ansari
ISLAM Malise Ruthven
ISLAMIC HISTORY Adam Silverstein
ISOTOPES Rob Ellam
ITALIAN LITERATURE Peter Hainsworth and David Robey
JESUS Richard Bauckham
JEWISH HISTORY David N. Myers
JOURNALISM Ian Hargreaves
JUDAISM Norman Solomon
JUNG Anthony Stevens
KABBALAH Joseph Dan
KAFKA Ritchie Robertson
KANT Roger Scruton
KEYNES Robert Skidelsky
KIERKEGAARD Patrick Gardiner
KNOWLEDGE Jennifer Nagel
THE KORAN Michael Cook
LAKES Warwick F. Vincent
LANDSCAPE ARCHITECTURE Ian H. Thompson
LANDSCAPES AND GEOMORPHOLOGY Andrew Goudie and Heather Viles
LANGUAGES Stephen R. Anderson
LATE ANTIQUITY Gillian Clark
LAW Raymond Wacks
THE LAWS OF THERMODYNAMICS Peter Atkins
LEADERSHIP Keith Grint
LEARNING Mark Haselgrove
LEIBNIZ Maria Rosa Antognazza
LIBERALISM Michael Freeden
LIGHT Ian Walmsley
LINCOLN Allen C. Guelzo
LINGUISTICS Peter Matthews
LITERARY THEORY Jonathan Culler
LOCKE John Dunn
LOGIC Graham Priest
LOVE Ronald de Sousa
MACHIAVELLI Quentin Skinner
MADNESS Andrew Scull
MAGIC Owen Davies
MAGNA CARTA Nicholas Vincent
MAGNETISM Stephen Blundell
MALTHUS Donald Winch
MAMMALS T. S. Kemp
MANAGEMENT John Hendry
MAO Delia Davin
MARINE BIOLOGY Philip V. Mladenov
THE MARQUIS DE SADE John Phillips
MARTIN LUTHER Scott H. Hendrix
MARTYRDOM Jolyon Mitchell
MARX Peter Singer
MATERIALS Christopher Hall
MATHEMATICS Timothy Gowers
THE MEANING OF LIFE Terry Eagleton
MEASUREMENT David Hand
MEDICAL ETHICS Tony Hope
MEDICAL LAW Charles Foster
MEDIEVAL BRITAIN John Gillingham and Ralph A. Griffiths
MEDIEVAL LITERATURE Elaine Treharne
MEDIEVAL PHILOSOPHY John Marenbon
MEMORY Jonathan K. Foster
METAPHYSICS Stephen Mumford
THE MEXICAN REVOLUTION Alan Knight
MICHAEL FARADAY Frank A. J. L. James
MICROBIOLOGY Nicholas P. Money
MICROECONOMICS Avinash Dixit
MICROSCOPY Terence Allen
THE MIDDLE AGES Miri Rubin
MILITARY JUSTICE Eugene R. Fidell
MILITARY STRATEGY Antulio J. Echevarria II
MINERALS David Vaughan
MIRACLES Yujin Nagasawa
MODERN ART David Cottington
MODERN CHINA Rana Mitter
MODERN DRAMA Kirsten E. Shepherd-Barr
MODERN FRANCE Vanessa R. Schwartz
MODERN INDIA Craig Jeffrey
MODERN IRELAND Senia Pašeta
MODERN ITALY Anna Cento Bull
MODERN JAPAN Christopher Goto-Jones

MODERN LATIN AMERICAN LITERATURE Roberto González Echevarría
MODERN WAR Richard English
MODERNISM Christopher Butler
MOLECULAR BIOLOGY Aysha Divan and Janice A. Royds
MOLECULES Philip Ball
MONASTICISM Stephen J. Davis
THE MONGOLS Morris Rossabi
MOONS David A. Rothery
MORMONISM Richard Lyman Bushman
MOUNTAINS Martin F. Price
MUHAMMAD Jonathan A. C. Brown
MULTICULTURALISM Ali Rattansi
MULTILINGUALISM John C. Maher
MUSIC Nicholas Cook
MYTH Robert A. Segal
THE NAPOLEONIC WARS Mike Rapport
NATIONALISM Steven Grosby
NAVIGATION Jim Bennett
NELSON MANDELA Elleke Boehmer
NEOLIBERALISM Manfred Steger and Ravi Roy
NETWORKS Guido Caldarelli and Michele Catanzaro
THE NEW TESTAMENT Luke Timothy Johnson
THE NEW TESTAMENT AS LITERATURE Kyle Keefer
NEWTON Robert Iliffe
NIETZSCHE Michael Tanner
NINETEENTH-CENTURY BRITAIN Christopher Harvie and H. C. G. Matthew
THE NORMAN CONQUEST George Garnett
NORTH AMERICAN INDIANS Theda Perdue and Michael D. Green
NORTHERN IRELAND Marc Mulholland
NOTHING Frank Close
NUCLEAR PHYSICS Frank Close
NUCLEAR POWER Maxwell Irvine
NUCLEAR WEAPONS Joseph M. Siracusa
NUMBERS Peter M. Higgins
NUTRITION David A. Bender
OBJECTIVITY Stephen Gaukroger
OCEANS Dorrik Stow
THE OLD TESTAMENT Michael D. Coogan
THE ORCHESTRA D. Kern Holoman
ORGANIC CHEMISTRY Graham Patrick
ORGANIZATIONS Mary Jo Hatch
PAGANISM Owen Davies
PAIN Rob Boddice
THE PALESTINIAN-ISRAELI CONFLICT Martin Bunton
PANDEMICS Christian W. McMillen
PARTICLE PHYSICS Frank Close
PAUL E. P. Sanders
PEACE Oliver P. Richmond
PENTECOSTALISM William K. Kay
PERCEPTION Brian Rogers
THE PERIODIC TABLE Eric R. Scerri
PHILOSOPHY Edward Craig
PHILOSOPHY IN THE ISLAMIC WORLD Peter Adamson
PHILOSOPHY OF LAW Raymond Wacks
PHILOSOPHY OF SCIENCE Samir Okasha
PHOTOGRAPHY Steve Edwards
PHYSICAL CHEMISTRY Peter Atkins
PILGRIMAGE Ian Reader
PLAGUE Paul Slack
PLANETS David A. Rothery
PLANTS Timothy Walker
PLATE TECTONICS Peter Molnar
PLATO Julia Annas
POLITICAL PHILOSOPHY David Miller
POLITICS Kenneth Minogue
POPULISM Cas Mudde and Cristóbal Rovira Kaltwasser
POSTCOLONIALISM Robert Young
POSTMODERNISM Christopher Butler
POSTSTRUCTURALISM Catherine Belsey
PREHISTORY Chris Gosden
PRESOCRATIC PHILOSOPHY Catherine Osborne
PRIVACY Raymond Wacks
PROBABILITY John Haigh
PROGRESSIVISM Walter Nugent
PROJECTS Andrew Davies

PROTESTANTISM Mark A. Noll
PSYCHIATRY Tom Burns
PSYCHOANALYSIS Daniel Pick
PSYCHOLOGY Gillian Butler and Freda McManus
PSYCHOTHERAPY Tom Burns and Eva Burns-Lundgren
PUBLIC ADMINISTRATION Stella Z. Theodoulou and Ravi K. Roy
PUBLIC HEALTH Virginia Berridge
PURITANISM Francis J. Bremer
THE QUAKERS Pink Dandelion
QUANTUM THEORY John Polkinghorne
RACISM Ali Rattansi
RADIOACTIVITY Claudio Tuniz
RASTAFARI Ennis B. Edmonds
THE REAGAN REVOLUTION Gil Troy
REALITY Jan Westerhoff
THE REFORMATION Peter Marshall
RELATIVITY Russell Stannard
RELIGION IN AMERICA Timothy Beal
THE RENAISSANCE Jerry Brotton
RENAISSANCE ART Geraldine A. Johnson
REVOLUTIONS Jack A. Goldstone
RHETORIC Richard Toye
RISK Baruch Fischhoff and John Kadvany
RITUAL Barry Stephenson
RIVERS Nick Middleton
ROBOTICS Alan Winfield
ROCKS Jan Zalasiewicz
ROMAN BRITAIN Peter Salway
THE ROMAN EMPIRE Christopher Kelly
THE ROMAN REPUBLIC David M. Gwynn
ROMANTICISM Michael Ferber
ROUSSEAU Robert Wokler
RUSSELL A. C. Grayling
RUSSIAN HISTORY Geoffrey Hosking
RUSSIAN LITERATURE Catriona Kelly
THE RUSSIAN REVOLUTION S. A. Smith
SAVANNAS Peter A. Furley
SCHIZOPHRENIA Chris Frith and Eve Johnstone
SCHOPENHAUER Christopher Janaway
SCIENCE AND RELIGION Thomas Dixon
SCIENCE FICTION David Seed
THE SCIENTIFIC REVOLUTION Lawrence M. Principe
SCOTLAND Rab Houston
SEXUALITY Véronique Mottier
SHAKESPEARE'S COMEDIES Bart van Es
SHAKESPEARE'S SONNETS AND POEMS Jonathan F. S. Post
SHAKESPEARE'S TRAGEDIES Stanley Wells
SIKHISM Eleanor Nesbitt
THE SILK ROAD James A. Millward
SLANG Jonathon Green
SLEEP Steven W. Lockley and Russell G. Foster
SOCIAL AND CULTURAL ANTHROPOLOGY John Monaghan and Peter Just
SOCIAL PSYCHOLOGY Richard J. Crisp
SOCIAL WORK Sally Holland and Jonathan Scourfield
SOCIALISM Michael Newman
SOCIOLINGUISTICS John Edwards
SOCIOLOGY Steve Bruce
SOCRATES C. C. W. Taylor
SOUND Mike Goldsmith
THE SOVIET UNION Stephen Lovell
THE SPANISH CIVIL WAR Helen Graham
SPANISH LITERATURE Jo Labanyi
SPINOZA Roger Scruton
SPIRITUALITY Philip Sheldrake
SPORT Mike Cronin
STARS Andrew King
STATISTICS David J. Hand
STEM CELLS Jonathan Slack
STRUCTURAL ENGINEERING David Blockley
STUART BRITAIN John Morrill
SUPERCONDUCTIVITY Stephen Blundell
SYMMETRY Ian Stewart
TAXATION Stephen Smith
TEETH Peter S. Ungar
TELESCOPES Geoff Cottrell
TERRORISM Charles Townshend
THEATRE Marvin Carlson
THEOLOGY David F. Ford

THINKING AND REASONING
Jonathan St B. T. Evans
THOMAS AQUINAS Fergus Kerr
THOUGHT Tim Bayne
TIBETAN BUDDHISM
Matthew T. Kapstein
TOCQUEVILLE Harvey C. Mansfield
TRAGEDY Adrian Poole
TRANSLATION Matthew Reynolds
THE TROJAN WAR Eric H. Cline
TRUST Katherine Hawley
THE TUDORS John Guy
TWENTIETH-CENTURY BRITAIN
Kenneth O. Morgan
THE UNITED NATIONS
Jussi M. Hanhimäki
THE U.S. CONGRESS Donald A. Ritchie
THE U.S. SUPREME COURT
Linda Greenhouse
UTILITARIANISM
Katarzyna de Lazari-Radek and
Peter Singer
UNIVERSITIES AND COLLEGES
David Palfreyman and Paul Temple
UTOPIANISM Lyman Tower Sargent
THE VIKINGS Julian Richards
VIRUSES Dorothy H. Crawford
VOLTAIRE Nicholas Cronk
WAR AND TECHNOLOGY
Alex Roland
WATER John Finney
WEATHER Storm Dunlop
THE WELFARE STATE David Garland
WILLIAM SHAKESPEARE Stanley Wells
WITCHCRAFT Malcolm Gaskill
WITTGENSTEIN A. C. Grayling
WORK Stephen Fineman
WORLD MUSIC Philip Bohlman
THE WORLD TRADE
ORGANIZATION Amrita Narlikar
WORLD WAR II Gerhard L. Weinberg
WRITING AND SCRIPT
Andrew Robinson
ZIONISM Michael Stanislawski

Available soon:

VETERINARY SCIENCE James Yeates
THE ENGLISH LANGUAGE
Simon Horobin
ORGANIZED CRIME
Georgios A. Antonopoulos and
Georgios Papanicolaou
THE PHILOSOPHY OF RELIGION
Tim Bayne
THE HELLENISTIC AGE
Peter Thonemann

For more information visit our website

www.oup.com/vsi/

J. L. Heilbron

THE HISTORY OF PHYSICS

A Very Short Introduction

Great Clarendon Street, Oxford, OX2 6DP,
United Kingdom

Oxford University Press is a department of the University of Oxford.
It furthers the University's objective of excellence in research, scholarship,
and education by publishing worldwide. Oxford is a registered trade mark of
Oxford University Press in the UK and in certain other countries

First published as *Physics: A Short History from Quintessence to Quarks* 2015
First published as a Very Short Introduction 2018

First edition published in 2018

Published in the United States of America by Oxford University Press
198 Madison Avenue, New York, NY 10016, United States of America

British Library Cataloguing in Publication Data
Data available

Library of Congress Control Number: 2017951924

ISBN 978–0–19–968412–0

Printed and bound by
CPI Group (UK) Ltd, Croydon, CR0 4YY

Contents

List of illustrations

Introduction: the Greek way

The Superconducting Super Collider (SSC), the dream of American high-energy physicists, would have had a circumference of 87 kilometres and a price tag to match. Its proponents justified the expenditure on several grounds. On the high ground, it would probe the universe to philosophical depths and thus 'keep faith with the Greeks'. Below ground, it would advance tunnelling technique and give society perfect sewers. Congress cancelled it in 1993.

The undertaking represented by the SSC might be the way to an ultimate physics. But it is not the Greek way. In antiquity, physics was philosophy, a liberal art, the pursuit of a free man wealthy enough to do what he wished. He did not aim to improve sewers and, since he had no need of public money, did not have to claim that he would. Nor did he want apparatus, since he seldom experimented; or mathematics, since he seldom calculated. The few ancient applications of mathematics to physics constituted a mixed science devoted to the description of phenomena rather than to the search for principles.

In the tripartite division of Greek philosophy, physics stood between logic and ethics. It inquired into the principles regulating the physical world from the high heavens to the Earth's centre, and from the human soul to the life of the least of living creatures.

It thus functioned as a necessary approach to ethics, or the principles of a good life. For two millennia the main practical value of physics lay in the ethical consequences of its versions of the way the world began and persists.

Greek physics, with its eye to ethics, its indifference to mathematics and experiment, and its independence of states and courts, is sufficiently distinct from an enterprise conducted by salaried teams requiring elaborate technologies and mathematical analyses to deserve a different name. Let it be *physica* and those who cultivated it *physici*. This short book describes some of the ways by which ancient *physica* became modern physics. It does not ransack history to find items in ancient and medieval science that look like physics, but sketches the place and purpose of *physica* in the societies that supported it. Hence the primary site(s) of cultivation receive special emphasis: the independent private school (antiquity), court and library (Islam), university (later Middle Ages), court again (Renaissance), academy (late 17th and 18th centuries), university again (modernity), and university-government-industry (postmodernity). Of course, successive forms did not annihilate their predecessors. Academies of science survive, primarily as honorific societies and depositories of history, although a few flourish as national channels for funding, consultation, and outreach. The scientific advisory apparatus of government may be considered the descendant of courtly science; and the Greek schools, with their characteristic discursive style, continue in the myriad seminars in which the world's nascent science is presented and anatomized.

Chapter 1
Invention in antiquity

Tradition follows Aristotle in identifying the earliest *physici* as some gentlemen of Miletus, and in specifying half a dozen other Greek speakers as their successors. In this philo-Hellenic creation myth, no Greek *physicus* learned anything from a barbarian during the 250 years between the times of the eldest Milesian, Thales, and Aristotle. The story that Pythagoras, if he existed, did so partly in Egypt, suggests outside input; and studies of cuneiform texts reveal a natural knowledge among the Babylonians in some ways more advanced than that of the ancient Greeks. Still, the essential criterion that Aristotle used to identify his predecessors was not that they were Greek, but that they had conquered a paralysing prejudice. Despite robust contrary evidence, they believed that the natural world runs on law-like principles discoverable by the human mind and immune from interruption or cancellation by meddling gods and demons.

This bold departure underlies and circumscribes all forms of *physica*, natural philosophy, and physics. Its implications go farther even than replacing caprice by law-like behaviour. Since the gods displayed all too faithfully the behaviour of human beings, de-deifying implied (to speak Greek) de-anthropomorphizing. The progress of physics has continued to remove human quirks and qualities projected onto nature. Thus nature, or the objective world, came to lose not only benevolence, malevolence, and colour, but also such apparently indispensable attributes as space, time, and causality.

Of the four main schools of ancient philosophy, Aristotle's paid greatest attention to *physica* (see Figure 1). Having a particular interest in zoology, he derived his fundamental principles with an eye to the classification of animals. Because of his emphasis on *physica* and because his philosophy dominated during the Middle Ages and beyond, convenience advises taking it as normative. In antiquity, however, it had to compete with Platonic, Epicurean, and Stoic philosophies. Typically, a school's premises, library, and other assets passed from the founder to his senior disciples, and, perhaps consequently, the schools bore names suggestive more of real estate than of scholarship: the Academy (Grove) for the Platonists; Lyceum (a shrine) and Parapatos (a place for walking) for the Aristotelians; the Stoa (Porch) for the Stoics; and the Garden of Epicurus.

During the 800 years from the founding of the Academy in Athens and its refounding in Alexandria, the four schools underwent many reversals of fortune. When operating normally, they were forums for free discussion of the sort practiced in political circles, but centred on studies pursued for personal improvement rather than for civic or financial advancement. Some students stayed with a single school for decades, others sampled each in turn. When Romans like Cicero frequented the schools of Athens they took advantage of this *Lernfreiheit*. The Pythagoreans did not form such a school, as they did not tolerate deviations from their doctrines and way of life.

Physica

Although *physica* ran from astronomy through zoology to psychology, limitation of coverage to cosmology and cosmogony, as in this book, is not unacceptably anachronistic, provided we recognize that the same principles of structure and change applied

1. School of Athens. Raphael's evocation of the intellectual vigor, diversity, and discursiveness of Greek Science.

to all natural processes. This truncated *physica* corresponds to the books of Aristotle dealing with general principles (*Physica*), the heavens (*De caelo*), the region between the Moon and the Earth (*Meteorologica*), and the creation and destruction of things on or in the Earth (*De generatione et corruptione*). These books (and the rest of the corpus from logic to ethics) became available in a standard format edited around 60 BCE from the lecture notes Aristotle bequeathed to his successors. They constitute the main parts of a theory of everything, or, as the moderns say, a TOE.

The first TOEs Aristotle stepped on belonged to the Milesian monists. He then took on more generous materialists: Leucippus and Democritus, who allowed two principles—atoms and the void; Empedocles, who accepted the three Milesian elements (water, air, fire) and added earth to complete the tetrad; and Anaxagoras, who admitted an infinite number of different sorts of stuff. There were also those whose matter had no stuff at all—Pythagoras notably, for whom number had an independent existence. The Pythagoreans' deduction that there must be a counter-Earth circulating opposite ours around a central fire (to raise the number of heavenly bodies to the holy tetractys) proved to Aristotle both the falsity of their *physica* and the nonsense to which numbers can lead.

Aristotle's teacher, Plato, inclined towards numerology. He took as his material not only mathematical abstractions but also supersensible idealizations of classes of objects: for example, the Idea 'Horse', in which individual horses 'participate' more or less, but always imperfectly. Consequently, although Plato was optimistic about the possibility of knowing the ideal world, and the supreme Good that made the Ideas and their relations intelligible, he did not allow the possibility of true knowledge of the things of this world. Since no material individual could express an Idea perfectly, our *physica* can never be other than fuzzy.

Physica comes from 'physis' meaning 'nature', which, according to Aristotle, 'is the source or cause of being moved or being at rest'. What makes things move? The early *physici* adumbrated four causes of change that Aristotle later codified. The monists and the atomists considered only the material cause. Empedocles and Anaxagoras provided action by taking some principles to be active and others passive—vague glimpses of efficient causes. Others saw the need to explain order in a universe of change, and hinted at a teleological or final cause, such as set by a cosmic Mind. And Plato supplied a fourth cause, the formal, the Idea in which a thing participates.

Aristotle's inventory of the cosmological ideas of his predecessors, including anticipations of the four causes of change, was not an idle retrospective. It confirmed that he had not overlooked anything fundamental. 'Of all who have discussed principles and causes none has spoken of any kind except those which have been distinguished in [my] discourses on Physics'. Aristotle's TOE, thus established as complete, makes use of some special concepts. A *substance* is any individual thing. The collection of its properties constitutes its *form*, which, contrary to a Platonic Idea, occurs only in union with *matter*. Form can be divided, although only mentally, into *essence*, which makes a substance the sort of thing it is, and *accidents*, which can change without causing the substance to alter its essence or kind. The essences of the four elements are easily stated: fire is dry, hot, and absolutely light; air is hot, moist, and relatively light; water is cold, moist, and relatively heavy; earth is cold, dry, and absolutely heavy. The elements can transform into one another, as fire evaporates water into air. Socrates may be warm or cold, but heated or cooled too much he will cease to be Socrates.

It may now be intelligible to state that a substance's matter and essence are its material and formal causes, and that the active qualities of hotness and moistness are the principal efficient causes of change. The final cause is the purpose for the existence of a form. Heavy bodies have gravity so as to fall towards the centre of the world, and light bodies have levity so as to rise towards the heavens, thus restoring order disrupted by the activity of animate creatures or the revolutions of the celestial spheres. These spheres and the stars and planets they carry cannot be made of the four elements, whose forms require that they move when unimpeded in a straight line towards or away from the world's centre. A fifth element, a *quintessence*, obeying its formal and final causes, circulates around the universal centre. With these few principles and some ad hoc adjustments, Aristotle worked his way from the Mind of the Unmoved Mover, which, as it can think only of the most sublime thing, can think only of itself, down to the mind of man, which, though no less self-centred than the Mind of the Universe, is changeable like everything else below the quintessential heavens. And where there is change, there cannot be certainty; the best a *physicus* can do is to find a 'rule [that] applies to what is always true or true for the most part'.

A few deductions from Aristotle's approximate *physica* that came under sustained scrutiny will give some impression of its general character. Every motion, whether change of place or colour or species, requires an external mover. In a vacuum there literally is no place (no reference material) by which a body can orient itself; hence, there can be no vacuum. The flight of an arrow implies a vortex in the air, which cedes a place to the tip while pushing in at the tail. The ambiguous role of the air, offering both resistance and propulsion, made an obvious difficulty. Another awkwardness arose from the absolute dichotomy between terrestrial and celestial physics. Because the heavens cannot change, transient phenomena that appear to take place there, like comets and meteors, must have their seat with lightning and the weather in sublunar regions.

Nothing, however, is more obvious than that the Sun influences the weather. How? Sometimes Aristotle wrote as if he thought that the Sun was hot, which would violate his proscription against terrestrial qualities in heaven. More often, he ascribed the seasonal powers of the Sun to its annual revolution, which, together with the rolling of the quintessential spheres, continually stirs up the sublunary regions. These disturbances cause moist and dry vapours to rise from the Earth. Precipitation results from the moist exhalation, winds from the dry. 'The same stuff is wind on the earth, and earthquakes under it, and in the clouds thunder'. Lightning and thunder are dry exhalations breaking free from the clouds. The rainbow is a reflection from the clouds. Aristotle uncharacteristically described it geometrically: the Sun, the eye of the observer, and the centre of the bow lie on a straight line that cannot exceed a certain angle with the horizon. This factoid would have a long and influential history.

Despite the continuing operation of final causes, the world is not in perfect order. The frictional drag of the turning lunar sphere on the stationary region of fire below it produces such anomalies as fiery meteors in the air, mountains above sea level, and water below earth. In the big picture, however, the universe resembles an onion. Peeled from the outside in, it discloses the fixed stars, the planets and luminaries in the conventional order Saturn, Jupiter, Mars, Sun, Venus, Mercury, and Moon, and, in slight disorder, the elements fire, air, water, and earth. What is outside the skin? Here the onion analogy fails. There is literally nothing there. And just as there is no space not included in the visible universe, there was no time at which it did not exist.

Aristotle's world picture thus lacked a creator. So did the atomic theory of Democritus, who, nevertheless, offered a creation story for the visible universe. It began to be when some atoms bouncing about in the infinite void chanced to form a great vortex. The largest fell to the centre, forming the Earth. From what remained, centrifugation produced the air, the luminaries, the planets, and the stars. Although the same dull stuff makes up everything, our sensory systems can build rich images from its few properties—the size, shape, and motion of its constituent atoms. Epicurus added a spontaneous 'swerve' to explain how atoms, falling in parallel through the void, occasionally collide and concatenate a world. According to him, the soul can exploit the swerve to choose to live the good life in an otherwise pointless universe. Since the Epicurean did not have to fear gods in this life or anything in the next, he could take moderate enjoyment of the flesh and free employment of the mind as the greatest goods. The inevitable erosion of all sound doctrine has transformed Epicurus' sober happiness into selfish hedonism.

Whereas the atomists allowed for the creation of many worlds in space and time by random accretion of their parts, the Stoics supposed that the single cosmos they admitted, geocentric like Aristotle's, is alternately destroyed and recreated. In place of disparate atoms, the Stoics put a continuous prime matter; and in place of bumps and grinds, 'pneuma', a self-moving elastic compound of fire and air that gives matter its cohesive and other properties. Strict causality applies everywhere, guaranteed and effected by the spatial continuity of the pneuma. The system would seem to rule out free will decisively. But since ethics required a free acceptance of fate and the mental preparation necessary to meet it, Stoics had to find a way around the strict causality of their physics. Their solution was no more plausible than the Epicurean swerve. As an alternative to strict atomism, however, the Stoic concept of a space-filling, active, elastic spirit had a future.

In contrast to the Peripatetics with their unchanging, uncreated cosmos, and the atomists and Stoics with their random and cyclical worlds, the Academics had a full cosmogony, with a creator as well as a creation story. As told by Plato's mouthpiece, the mathematician Timaeus of Locris, the Demiurge who made the realm of Ideas used what he had left after rolling out the celestial equator and the paths of the planets to manufacture human souls. These he sent to stars in the realm of Ideas to await planting in bodies created by the lesser gods to whom he assigned the task of making the sensible world. At the end of life, the rational soul returns to its star if its human possessor has lived a good life; if not, the soul reincarnates in a lesser being. Our animal parts serve merely to keep our head, the seat of our reason, from rolling around on the ground.

Used properly, our rational soul can bring us via observation of the motions of the heavenly bodies to the discovery of number, time, and harmony, and to the contemplation of the Ideas. We might then perceive that the Ideas of the four elements and the quintessence are linked to the mathematics of the five regular solids. The plane faces of three of them (the tetra, octa, and icosahedron) are equilateral triangles, and consequently the elements corresponding to them (fire, air, and water) are interconvertible (see Figure 2). The remaining two, the cube and the dodecahedron, are the 'Ideas' of the Earth and the universe as a whole. We should not press the obvious difficulties. The lesser gods who created the material world were not entirely competent. '[I]t is fitting that we should, in these matters, accept the likely story and look for nothing further'.

Roman worlds

Although foreign students from Rome added little of note to Greek *physica*, they turned much of it into useful summaries and compendia. During the last years of the Republic, Lucretius versified Epicurus' atomism. During the reign of Caesar Augustus, the poet Ovid made a conspicuous place for Pythagoreanism in his *Metamorphoses*. Under the Emperor Nero, Seneca composed a meteorology on Stoic principles—cut short, unfortunately, by Nero's invitation (which Seneca could scarcely refuse) to commit suicide. And in the years before his fatal inspection of Vesuvius during its eruption of 79 CE, Pliny the Elder crammed into one of the thirty-seven books of his *Natural History* a qualitative survey of the world featuring a God unpolluted by commerce with human beings; a naturalistic account of meteors, comets, and eclipses calculated to free humankind from fear of lesser gods; some hints

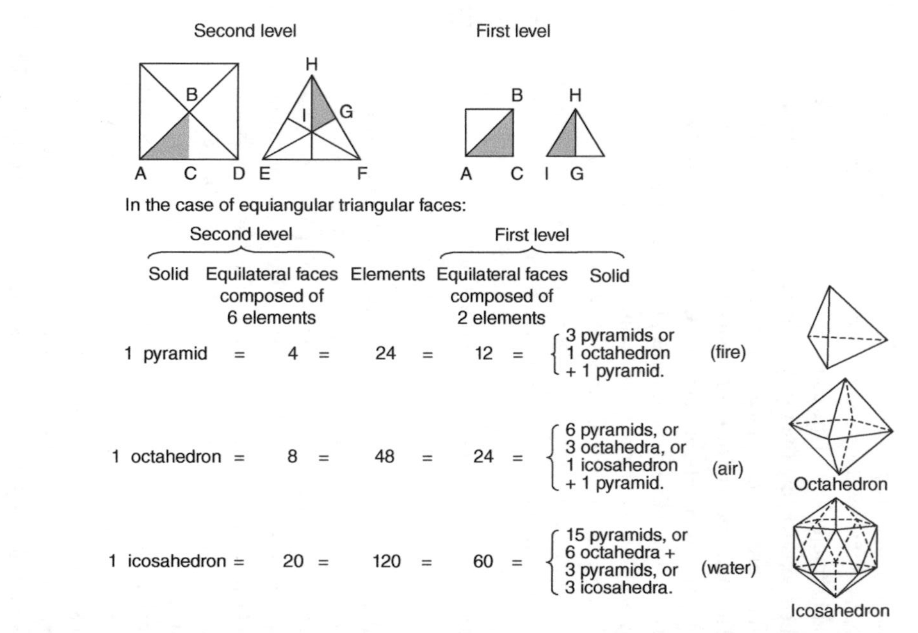

2. Geometrical foundations of the world. The faces of the cube come in aggregates of elementary isosceles right triangles ACB; those of the pyramid, octahedron, and icosahedron come in aggregates of various sizes of elementary 30-60-90 degree triangles IGH. The diagram shows how the second level aggregates from the first and, in the case of the three solids with equilateral triangles as faces, how the solids can transform into one another.

at atomism and stoicism (*physica* of chance and necessity); and a few bars from Pythagorean music of the spheres.

Lucretius begins his poem with the customary invocation of a muse, in this case Venus, after which he disobligingly announces that his great purpose is to remove her and all the other gods from human concerns. The atoms cavorting in the infinite void follow the laws of necessity apart from the occasional unintelligible swerve that gives spontaneity to the world and freedom to the will. No external agency can interfere with this process. Though thou be nothing but a chance congeries of buzzing particles, be joyful, submit to fate, and fear not the impotent gods.

The exposition of Pythagorean doctrine that Ovid jams into his catalogue of beings that transform into beasts treats only briefly the founder's *physica*, his teachings about

> The great world's origin, the cause of things | what nature is, what god, and whence the snow | what makes the lightning, whether thunder comes | from Jove or from the winds when clouds burst wide | why the earth quakes, what ordnance controls | the courses of the stars.

Most of Ovid's account of Pythagorean thought concerns metempsychosis and vegetarianism, a belief and a practice indissolubly connected. If souls transmigrate, how can you tell whether your ox, your faithful brother at the plow, was not in fact your late brother?

This sound teaching fell on deaf ears. Ovid admitted as much; Seneca, writing half a century later, reported that would-be Pythagoreans could not find a teacher. But that, according to Seneca, merely mirrored the sad state to which philosophy in general had fallen: 'Many philosophical lineages are dying out without a successor.' His *Natural Questions*, an attempt to reinvigorate Stoic *physica*, reconstructed the meteorology of the

Stoa and the Lyceum to answer the questions whose solution Ovid credited to Pythagoras—the causes of the wind and weather, thunder and lightning, and earthquakes. These had been standard problems in *physica* since the time of Thales. In Seneca's meteorology, high winds, thunder, and earthquakes are explosions of previously constrained pneuma. Breaking from Stoic authority, he enrolled comets, famous for bad reputations, among the planets and toyed with the possibility that their daily motions, and those of the stars in general, arise from a revolution of the Earth rather than from the turning of the celestial vault. As his placement of comets suggests, the interconnectedness of the Stoics' space-filling active pneuma destroyed the barrier between the heavens and the Earth characteristic of Aristotelian cosmology.

Seneca puffed up his meteorology with moral reflections that appear to be its reason and result. If you do not enquire into the material foundation of the world and the nature of its creator or guardian, or speculate whether he still creates (if he ever did) or has retired, whether he thinks only of himself, and whether he can amend fate, you enslave yourself to merely human affairs. Life would not be worth its pain and suffering were it not for the opportunity to learn that nature, fate, the world, providence, and God are different names for the same thing; that God did not make the world exclusively for human beings; that life is a sentence of death and philosophy, by dispelling fear of death, is the only healthy way to resign oneself to fate. Seneca's eloquent combination of moralizing and meteorology made the *Natural Questions* the authority on the physical problems it treated for most of the Latin Middle Ages.

Had Plutarch's dialogue on the markings on the Moon not suffered an eclipse of fifteen centuries, Seneca's stoicism would have faced a sprightly challenge in medieval times. Plutarch proposed a question that embarrassed several ancient systems: why do we not see the Sun's image on the Moon as we see it on the sea? Since the answer depends on knowledge of the Moon's makeup, mathematicians could not answer it. What then did the *physici* have to say? The mottled lunar surface shows that the Peripatetic model of the Moon as pure quintessence is nonsense. The Stoic view—that it is made of a sort of *pneuma*—is worse, since air and fire cannot reflect visual rays. Plato's model, which gives the Moon a rocky surface that acts as if composed of a great many randomly oriented mirrors, is the best option.

What then keeps the rocky Moon from falling on the Earth? '[T]he rapidity of its revolution, just as missiles placed in slings are kept from falling by being whirled around in a circle'. What keeps it close to Earth? A literal law of nature: 'the position of earth lays an action against the Moon and she is legally assignable by right of propinquity and kinship to the earth's real and personal property'. Is the Moon inhabited? Very likely, but by creatures as different from us as we are from fish, and also by disembodied souls awaiting their next incarnation. Plutarch's ideas about the Moon, minus his metempsychosis, recur in Galileo and Kepler.

During the philosophical downturn mentioned by Seneca, the Platonic Academy did not exist. It had come to a temporary end during the Roman conquest of Athens in 86 BCE after a period of intense scepticism about the possibility of achieving any secure knowledge about anything. It revived in 410 CE with a curriculum based on 'Neoplatonism', invented in Rome by Plotinus, who died in 270 CE, and systematized by his disciple, Porphyry. Plotinus was not a sceptic. His confident speculations rose above even the Demiurge, whose employment as creator, even if only of the rational realm and the lesser gods, seemed to him incompatible with occupying the acme of the divine pyramid. How did the Demiurge relate to the impassable God or Good that Plato had put in charge of everything? Plotinus answered that God stands immovable at the head of a chain of created and creating 'emanations'. The first of these emanations, Intellect, contains the Ideas; the second, Soul, contains Nature; the next is the Demiurge. Plato had not supplied much information about the natural world created by the Demiurge's collaborators, and his thesis that we cannot know such things in principle inhibited elaboration of Timaeus' hints about polyhedra. So Neoplatonism grafted on Aristotle's *physica* as its account of the visible universe created by the demigods employed by the Demiurge.

Another conflation assimilated the top Neoplatonic powers—the One, the Intellect, and the World Soul—with the Christian Trinity. In orthodox Christianity, God did not employ a vicar in his creative works. But several Christian sects still competitive in the 4th century taught that God the Father created only spiritual beings, one of whom, Jehovah, broke the chain of spiritual emanations and materialized his successors. Neoplatonism was the leading philosophy, and its co-inventor Porphyry the most effective critic, of Christianity during the late empire. His opponents, the architects of Christianity concerned to achieve standardization of belief, faced the challenges of harmonizing discrepancies among the four canonical gospels, explaining evil in the creation of a beneficent deity, and defining the relationships among the Persons of the Trinity. The most abstruse of their conundrums probably could not have been concocted without the language and concepts of Aristotelian *physica* and Neoplatonic philosophy.

Applications

The speculations of *Physici* were less useful to princes than the advice of applied mathematicians. The greatest of them all, Archimedes, worked for a tyrant of Syracuse. Although best known now for his detection of a counterfeit crown and his legendary feats in defence of Syracuse, Archimedes preferred to be remembered in the manner of a *physicus*, as a liberal artist. According to Plutarch, he regarded his practical accomplishments as 'mere accessories of geometry practiced for amusement' and 'every act that ministers to the needs of life as ignoble and vulgar'. Plutarch added that Plato had chastised two clever mathematicians, Eudoxus and Archytas, for turning their hands to practical things, and becoming 'corruptors and destroyers of the pure excellence of geometry'.

In contrast to the many useful applications of geometry, *physica* could do little more for the practical man than veneer him. Vitruvius recommends in his *Ten Books of Architecture*, which dates from the early Roman Empire, that the student learn *physica* to be able to judge excellence, to achieve authority, and to grow 'courteous, just, and honest'. More obviously, knowledge of 'the principles of physics [as taught] in philosophy' was needed for constructing waterworks; of the principles of musical intervals, for tuning catapults; of astronomy, for choosing sites for buildings; of meteorology, for avoiding places exposed to winds, lightning, and earthquakes. None of this related to the main business of construction, and for good reason. *Physica* knew nothing about the strength of materials. When it came to foundations, Vitruvius could only advise to 'dig down . . . as deep as the magnitude of the proposed work seems to require'.

A more lasting application of physical principles occurs in the writings of Ptolemy of Alexandria. He distinguished his astronomy, enshrined in the thirteen books of his *Syntaxis mathematica* or *Almagest* (as it is usually named after its Arabic translation), from his four books of astrological interpretation, the *Tetrabiblos*. The *Almagest* considers planets only as moving points; the *Tetrabiblos* ascribes to them the active elemental qualities that bring their influence down to Earth. The planets are warm, moist, cold, and dry in different degrees depending upon their distances from warm Sun, moist Moon, and cold and dry Earth. Judged by temperature and humidity, Jupiter, Venus, and Moon are beneficent; Mars and Saturn evildoers; and Sun and Mercury ambiguous. From this astrophysics, Ptolemy derived a physical anthropology that explained why Ethiopians are black; Scythians white; inhabitants of the temperate region medium in colour, civilized, and sagacious; and himself mathematical.

Although the astrophysics of the *Tetrabiblos* admitted earthly qualities into the celestial regions, Ptolemy's astronomy revolved in the quintessence of the Aristotelian universe. According to Aristotelian celestial mechanics, each planet and luminary travels on a sphere concentric with the Earth. But none of the Peripatetics, including that Eudoxus whom Plato criticized for doing carpentry instead of geometry, could 'save the phenomena' consistently with the concentric principle. The obvious explanation of the notable alterations in brightness of the Moon and the planets is their changing distance from Earth. So mathematicians dismissed Aristotle's spheres in favour of non-concentric circles on which to mount celestial objects.

Ptolemy's *Almagest* saves the phenomena by displacing Earth from the centre of the Sun's circular orbit, thus representing phenomena arising from the ellipticity of Earth's orbit as an effect of perspective (see Figure 3a). Moon and planets likewise require circles ('eccentrics') with centres displaced, though in different directions and magnitudes, from Earth's. They also require a second circle ('epicycle') to model the effects of Earth's revolution around the Sun, during which a 'superior' planet (Mars, Jupiter, or Saturn) appears to reverse directions periodically in its orbit around Earth (see Figure 3b). A splendid refinement, mimicking the effects of ellipticity almost perfectly, made the centre of the epicycle rotate with constant velocity around an 'equant' point placed as far on one side of the centre of the eccentric as the Earth was on the other (see Figure 3c).

This brilliant, complex bric-a-brac violated good physics by postulating revolutions around unoccupied points without giving any physical reason for their position or motion. They were fictions introduced for description, not explanation. Ptolemy was not content with the division of labour by which *physici* aimed at the truth about substance and mathematicians at description of its accidents. Reversing the usual precedence, he declared the priority of mathematics over physics in the *Almagest* and

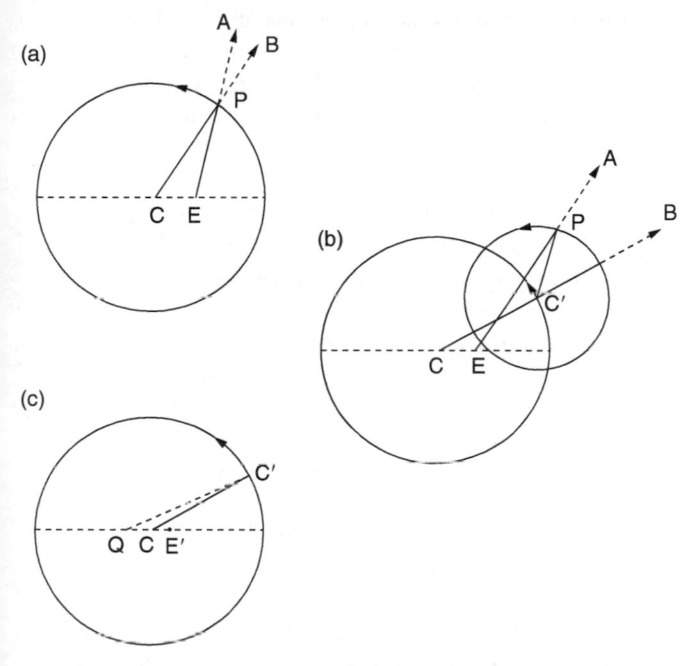

3. Ptolemaic bric-a-brac. (a) Saving the first anomaly: P, the planet orbiting at constant velocity on a circle centered at C ('the eccentric'); EA, CB, directions of P as seen from E and C, respectively. (b) Saving the second anomaly: C′ , the center of P's epicycle, moves cc (counterclockwise) on eccentric C as P moves cc around it, both motions at constant velocity. (c) The equant: C′ moves cc with constant velocity around Q ('the equant point') where QC = CE′ = CE/2.

exemplified it in the astrophysics of his *Hypotheses of the Planets*. Here he made the Sun run in a groove between two concentric spheres whose common centre is offset from the centre of the Earth. Each superior planet rides on a marble (a reified epicycle) running in a similar groove between another pair of eccentric spheres. Accepting the Aristotelian (and Stoic) prohibition of a vacuum, Ptolemy close-packed his system so that the apogee of one celestial body coincides with the perigee of the next, and the greatest distance of Saturn lies in the vault of the stars. Knowing from eclipse observations the average lunar distance in terms of the Earth's radius r and, from his astronomical calculations, the ratios of the radii of the epicycles and eccentrics of each planet in his close-packed system, Ptolemy worked out the distance to the stars to be just shy of $20{,}000r$.

This number, though vastly less than the true radius of Saturn's orbit, was large enough to suggest the insignificance of humankind, especially when expressed in miles. Eratosthenes of Cyrene, one-time head of the Library of Alexandria, converted r to human measure by determining the angle α between the vertical and the direction of the Sun's rays at Alexandria at noon at the same instant that it stood overhead at Aswan, 5,000 stadia due south. Since α equals the difference in latitude between the two places, Eratosthenes could calculate, from $2\pi r{:}5{,}000 = 360°{:}\alpha$, that $r \approx 4{,}000$ miles (see Chapter 2, Figure 8). The radius of the visible world would then be eighty million miles. These numbers were scarcely bettered before the 17th century.

Among other enduring pieces of mixed mathematics from the Hellenistic period are Archimedes' demonstration of the law of the balance, his hydrostatics with its associated concept of specific gravity, an anonymous treatise on mechanics incorrectly attributed to Aristotle, and parts of texts on optics by Euclid and Ptolemy. The optical works, although mainly geometry, presuppose at least three physical principles of interest: the rays responsible for vision originate *from* the eye, change direction abruptly at the surface between media of different density, and take the shortest path between eye and object. Thus the angles of incidence and reflection are equal. A treatise perhaps written by Ptolemy offers measurements of the breaking of visual rays when they strike an air–water surface. The method gave results that differ insignificantly from Snell's law for angles of incidence up to 40°. The Ptolemaic author observed that refraction must also occur at the boundary between air and ether (or fire or quintessence) and, although he could not calculate the amount, rightly stated that the effect makes celestial objects near the horizon look higher in the sky than they are.

The pseudo-Aristotelian *Mechanica* exceeds the scope of *physica* in being addressed to mechanical problems. But it begins with some physics talk about the paradoxical marvels of circular motion. In a spinning wheel, the highest point moves forward and downward as the lowest point moves backward and upward; and the rim, though larger than the hub, moves faster. Pseudo-Aristotle explains that the shorter the spoke, the closer its extremity to the stationary centre. By an assumed principle of continuity, the proximity inhibits the motion. On the principle that longer radii are moved more easily through a given angle than shorter ones, the author delivers the law of the lever and deftly applies it to oars, rudders, sails, wheels, pulleys, capstans, wedges, tooth extractors, and nutcrackers. From the nutcracker he proceeds to the headcracker later known as 'Aristotle's wheel'. Two circles fixed together concentrically roll horizontally. Although the larger lays down a longer path than the smaller, they stay together. Why? 'It is strange... and wonderful'. The wonder rolled all the way to Galileo, who took many hints and problems from the *Mechanica* believing that they came from Aristotle.

Ancient applications of what today would be deemed engineering physics included the hydraulics of the magnificent system of aqueducts, tunnels, and fountains that supplied Rome's water. 'There was never any desseine in the whole world enterprised and effected, more admirable than this', so said Pliny, the expert on marvels. The ancients also knew of the suction pump and waterwheel but made little use of either. Perhaps the best known of their waterworks apart from aqueducts push water around for theatrical effects. The greatest impresario in this line, the military engineer, Hero of Alexandria, opened his treatise on ingenious devices with proofs that air is corporeal and that, in union with the other elements and principles, it could amaze the most jaded.

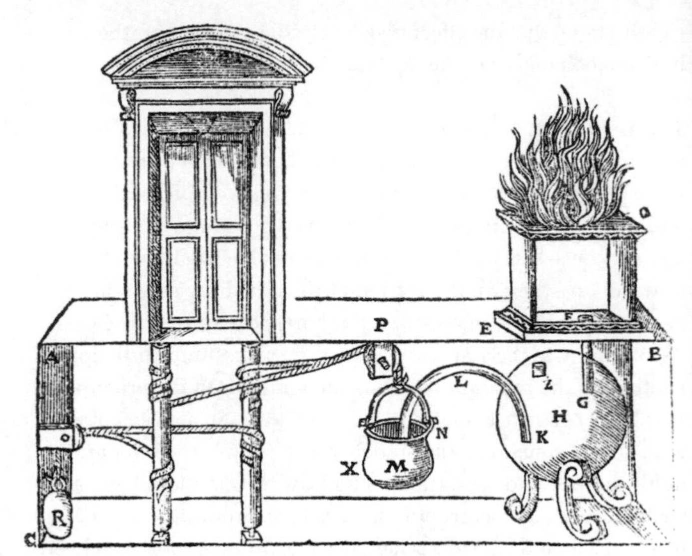

4. Sabbath machinery. Air heated by the fire on the altar pushes water and steam from the vessel H into the pot M, which, in descending, pulls ropes that cause the axles they encircle to rotate and open the temple door.

For example, concealed hollow tubes running inside a wine-filled altar brought wine to the altar when a fire was kindled on it. Another mechanism allowed the sacred fire to open temple doors (see Figure 4). Such games did not lead to steam engines.

Dumbing down

Owing to lack of interest, disturbances, or the rise of Christianity, Latin speakers gradually ceased frequenting Greek schools and cultivated Greek slaves, who taught their owners their language, became scarce. Knowledge of Greek, and the copying of Greek texts, dwindled, to arrive functionally at zero by the time the Ostrogoth Theodoric became King of Italy in 493. Curiously, Theodoric knew more Greek than his learned Latin subjects, having been reared as a hostage in Byzantium.

Christian education beyond the Latin schools that taught skills useful to orators and bureaucrats was haphazard and, in the higher reaches of biblical interpretation, Christians were largely self-taught. That might help account for the breadth of opinion, and frequency of heresy, among the early self-made theologians. Many of them conveyed snippets of the *physica* they knew in their Bible commentaries, particularly on the six days of creation (Hexaemeron). Unfortunately, Augustine's call in his *De Doctrina Christiana* for a dictionary of ancient science to assist understanding of Scripture met with only partial success.

Lacking such a text in the 5th century, Latin admirers of *physica* resorted to the many compendia copied or composed during the time of troubles. Of first importance was Pliny's *Natural History*. Drawing on him and others, Macrobius Theodosius, perhaps a Greek-speaker in Theodoric's service, sugarcoated the liberal arts in a *Commentary on the Dream of Scipio* that would suit the taste of the Middle Ages. In this dream, Cicero describes a visit of the deceased Scipio Africanus, the scourge of Carthage, to Scipio's grandson. Africanus reports that the souls of true statesmen dwell in the Milky Way between incarnations, and he transports young Scipio there to compare the vastness of the heavens and their harmonious revolutions with the pettiness and discord of the greatest terrestrial empires.

Proceeding downwards from the soul emporium of the Milky Way, Macrobius defines the usual reference circles of astronomers (equator, ecliptic, and so on), describes the apparent motions of Sun, Moon, and planets; and surveys Earth's five climes, impenetrable torrid zone, and surrounding ocean. Earth is so small a speck compared with the heavens that the pressure of the surrounding air easily sustains it in the world's centre. Thus suspended, its inhabitants can descry that the Demiurge gave the planets and luminaries distances and periods keyed to the intervals Pythagoras had discovered by plucking lute strings on Earth. If Macrobius knew that these intervals differed altogether from the planetary parameters calculated by Ptolemy, the discrepancy did not bother him. He referred quantitative details to people 'disengaged from serious matters'.

While descending from the Milky Way to its temporary abode on Earth, the pure soul acquires reason from Saturn's sphere, power to act from Jupiter's, boldness from Mars', sense perception and imagination from the Sun's, passion from Venus', speech and analytical capacity from Mercury's, and, from the Moon's, the ability to perform the physical functions incident to life among the dregs of creation. Macrobius conveys much interesting misinformation in describing this journey: for example, that the planets shine by their own light, the Earth does not reflect sunlight, and Venus and Mercury lie between the Sun and Mars. Giving Plotinus as his authority, Macrobius makes the planets and luminaries merely the signs, not the movers, of future events.

Macrobius' blend of *physica*, Neoplatonism, numerology, metempsychosis, moralizing, and misinformation was not the only vehicle of transmission of ancient texts produced by high-level civil servants (if such he was) at the court of Theodoric. Theodoric's chief advisor, the Catholic Boethius, knew Greek. Concerned that his employer's friends and enemies threatened the survival of philosophy, he undertook to translate all of Aristotle's works. He began with the *Organon* (Aristotle's logical treatises), which, being useful in religious controversy, served as a Trojan horse for the rest of Peripatetic philosophy. Boethius did not complete much of his project, however, before Theodoric, believing him a traitor, had him killed. While awaiting execution, Boethius wrote his famous *Consolations of Philosophy* grounded on a Christian Neoplatonic world view that supported his Stoic resolve. Together with his *Consolations* and Aristotle's logic, Boethius bequeathed to medieval letters a short tract on music, which covered the terrestrial and celestial types, and the harmony of the soul. His successor as Theodoric's minister was a lawyer, Cassiodorus, who also laboured to preserve ancient learning for Christians. He set up a monastery for whose inmates he rendered the seven liberal arts in a manner suitable to Christians struggling to preserve some remnants of Roman culture in an increasingly barbarous world.

The most popular rendition of the seven liberal arts bequeathed to the Latin Middle Ages was a work of the 5th century, the *Wedding of Mercury and Philology*. Its author, Martianus Capella, set forth what little he knew about the liberal arts in lectures by philology's bridesmaids. Much of what they say is unintelligible. Thus Astronomy, having stepped from her ball of heavenly light, declared that during the 40,000 years she spent in Egypt studying her subject she learned that celestial natures, 'circling by their own surging, are diffused the entire way around in globular belts and circles'. After this venture into *physica* she described the apparent motions of the planets and luminaries, defined the equator and zodiac, tried to explain how to measure the Earth, climbed back into her ball, and, like the antiquity she represented, flew away.

Chapter 2
Selection in Islam

Among the Christians who equipped themselves with Aristotle's *Organon* to dispute about the nature of the Trinity and other enigmas were followers of the patriarch of Constantinople, Nestorius. After his excommunication in 431 CE, Nestorians settled in Persia, at Jundishapur, a town built 200 years earlier to house captive Roman soldiers. There they built an important school and hospital in which they enlarged their study of Aristotle and Galen. Vigorous missionaries, they became experts in translating difficult texts between unrelated languages. The early Arabian conquests to the east increased the scope of Nestorian missionary activity. The caliphs usually considered them as allies and the Qur'ān recognized them as fellow 'People of the Book', provided they paid their taxes.

The first era of Arabian expansion ended in the mid-8th century. The dynasty that had presided over most of it, the Umayyads, headquartered in Damascus, succumbed to a clan with Persian backing, the Abbasids, who built a new city, Baghdad, as its power base. Nestorian learning found a growing audience in the eclectic urban elite drawn to the new capital. In 765 the director of the hospital at Jundishapur became the Abbasids' court physician. Syriac-speaking Christian doctors soon swarmed in the new capital. They began the process of making Aristotle 'The Philosopher' in the Arabic language.

The project to render Greek science into Arabic lasted three centuries. The translators had to invent technical terms, even a word for 'philosophy', for which they adopted the transliteration *falsafa*. In the early 9th century, the Abbasid Caliph al-Mamūn established an academy or research centre and library at Baghdad, the House of Wisdom, which gathered works by Greek medical writers and mathematicians as well as Aristotle's. By 1050 every Aristotelian

5. Lecture in a House of Wisdom. The site is the library in Basra, the time the 13th century.

treatise except the *Politics* existed in Arabic, together with many books like the *Mechanica* incorrectly ascribed to him (see Figure 5).

A wide competitive search for ancient manuscripts marked the heyday of the House of Wisdom. This unusual treasure hunt had to do with the rapid rise of the Islamic state. Lacking administrative experience, the Arabian conquerors employed the *dīwān*, or bureaucratic offices and procedures of the Persian Empire, to tax, survey, and record. The new empire kept its books in Persian until late Umayyad times when it shifted to Arabic. That required translation of the *dīwān* manuals, which probably included introductions to arithmetic, geometry, and branches of applied mathematics.

The shift to Arabic opened the bureaucracy to place hunters. Mastery of applied mathematics may have helped people with ambitions of upper-level appointments; hence the hunt for mathematical treatises in depositories in Byzantium and elsewhere. Once domesticated by this process, mixed mathematics became a pursuit in its own right, though still supported for its applications to administration, engineering, and religious observance. Thus, unlike the Greeks, most Muslim *physici* and mathematicians were courtiers, officials, or functionaries.

The arts of reasoning and persuasion, useful for making headway in Baghdad's sophisticated philosophical salons and a name in religious controversy, again served as a stalking horse for the rest of Aristotle's system. The early Abbasids favoured a liberal form of *kalām* (scholastic exegesis of religious writings), *Mu'tazila*, which allowed wide scope to reason in the interpretation of Scripture and in the attainment of moral as well as philosophical truth. The Aristotelian corpus showed what reason unaided by revelation could achieve. It also bore directly on such practical matters as astrology and meteorology, and its vocabulary provided the means for discussion, if not for solution, of most questions about everyday experience.

These practical considerations may be taken as the efficient cause of the translations, and of the commentaries and compendia descended from them. Principles of knowledge and ways to wisdom might be adduced as formal and final causes. The material cause was paper. The Arabs became acquainted with this Chinese invention during their conquests in Central Asia. A paper mill operated in Baghdad before 800. A large market in books developed, supplied by an assembly line, or rather circle, of scribes who simultaneously took dictation from an author or scholar. The scribes read back what they had taken down and the dictator, if satisfied, 'authorized' the copies. The system produced more—and more faithful—versions of the originals than monastic scribes could make one at a time on vellum or parchment. Colossal libraries came into being, the largest, at perhaps 500,000 volumes, being over a hundred times the size of their nearest competitors in the West.

Central administration of the new empire proved impossible. Once the great expansion peaked in the 8th century its parts began to separate. In the 10th century, caliphates split off in Egypt under a dynasty claiming direct descent from the Prophet's daughter Fatima, and in Andalusia under a branch of the Umayyads. At the same time, the Abbasids lost all power in the East to Turkish and Persian shahs and emirs. These arrangements proved fleeting: in 1031 the Umayyads lost Andalusia to fundamentalist Muslims from North Africa; in 1171 the Fatimids gave way to the Kurd Saladin; in 1258 the Mongols destroyed Baghdad. Nonetheless, a common language for literature and administration, and conversions to Islam, kept the disparate political elements loosely together.

Because of the wide distribution in place and time of niches for its cultivation, *falsafa* could develop only in fits and starts in Islamic lands. Hence the enormous energy devoted to producing digests, encyclopedias, and commentaries on standard material by the most gifted of the Islamic philosophers, and a reason that they did not develop a *physica* much different from what they inherited. Another reason is that, in contrast to mathematics, *physica* had consequences for faith. Aristotle's Indifferent God, whether pure or Neoplatonized, could be Islamized, as he was Christianized, only at the expense of the logical consistency of the system. To insist on pure Aristotle, with his eternal world and perishable soul, invited theological opposition.

Falsafa

The House of Wisdom established in Baghdad by al-Mamūn incorporated the library of his father, Harun al-Rashid, the Caliph of the *Thousand and One Nights*, and an observatory for other nocturnal activities. It also sponsored expeditions to measure the size of Earth. Like the translation project, the astronomical and geographical observations arose from Greek prompts. Al-Mamūn wanted to check Ptolemy and Eratosthenes for the surer estimate of the size of his empire and the positions of the principal places in it. Geography flourished wherever encouraged and, with history, made the largest body of literature in the Islamic languages during the Middle Ages, apart from religion.

The great figure of the House of Wisdom, al-Kindī, the founder of Arabic Aristotelianism, is a good example of the generalization made earlier about the social status of Muslim savants. His father, from Yemen, was governor of Kufa in Iraq, whence al-Kindī made his way to Baghdad. In his version of Aristotle, the visible world consists of the four restless elements, the lazy circulating quintessence, form, matter, substance, and accident, and runs on the four causes. The Unmoved Mover transmits motion, and brings about substantial change, through the rotation of the heavens. This stationary Being is Neoplatonic in holding in its mind the Ideas and Forms of created things and Qur'ānic in knowing the particulars of the physical world and the characters and deeds of everyone alive and dead. With this adjustment, al-Kindī's world picture became the standard model of Aristotelian *physica* until the 12th century, when defenders of *falsafa* in Spain advocated a stricter reading of the Philosopher.

At first the amalgam fared well, especially in the hands of al-Fārābī, born the son of a Muslim general in remote Turkestan. He arrived in Baghdad after schooling by a Nestorian Christian to become the 'Second Teacher', Aristotle being the first. His ideal syllabus, *The Attainment of Happiness*, advises the truth seeker to begin with things easiest to understand—numbers and geometrical figures—and proceed gradually towards the material world. The journey leads through optics, astronomy, music, and mechanical principles taken as Archimedean abstractions to *physica*, the science of the things that make up the world. Proceeding again from the least material to the most, the pursuit of happiness sets out from the heavenly bodies and descends through the four elements to stones and Earth's interior.

The inquisitive mind next inquires about beings more perfect than nature and natural things, and sublimes to metaphysics. Beginning with the soul and intellect of the rational animal, it discovers the way of human perfection and climbs back up the chain of principles, peeling away the material aspects, ascending at last to the First Principle, the Being too perfect for description. The fulfilled mind now understands the nature and place of everything in the universe, including the Islamic state: the caliph relates to his hierarchy of subordinates as the First Principle relates to created beings from the highest intellect to the densest stone.

This clear teaching and Abbasid support did not suffice to establish a new Lyceum in Baghdad. Neither al-Kindī nor al-Fārābī led continuous academic lives there. Al-Mamūn's effort to impose *Mu'tazila* failed and his successors oppressed its representatives. The House of Wisdom almost perished. Al-Kindī had to leave. Al-Fārābī retained the support of his caliph, with the consequence that they were driven out of Baghdad together. After the Second Teacher died eight years later in Damascus, *falsafa* moved far into Persia, where in 980 it produced one of its finest flowers, Ibn Sīnā (Avicenna), in Bukhara, in Uzbekistan. Extravagantly precocious, Avicenna drifted around Persia in search of knowledge, often serving as a physician to the reigning power. His capacious memory held the entire Qur'ān, and much of Greek medicine, mathematics, and *physica*, which he poured into an antidote to ignorance and insomnia entitled *The Cure*.

The Cure proposes the same project as al-Fārābī's *Attainment of Happiness*, but arranges the steps in the order of the Aristotelian canon. Thus after the *Organon* come the basic principles of *physica*, their applications to the celestial and sublunar realms, the mysteries of motion and the human mind, and then mathematics. The last topic is First Philosophy, culminating in knowledge of the One who operates in the Neoplatonic manner of emanation and delegation. Seeking the unification of all knowledge sanctioned by the Qur'ān's insistence on the oneness of God, Avicenna took on human activities (prayer, prophecy, politics, law) as well as the natural and supernatural worlds.

Avicenna's widely read works ended, or suspended, the productive cultivation of *falsafa* in the eastern reaches of Islam. That was owing primarily to al-Ghazālī (Algazel), a conscientious theologian well versed in *kalām* and *falsafa*. After pondering the relations between reason and revelation, he announced that *kalām* could resolve nothing important for faith, and that *falsafa* was inimical to it. He pointed out that the Unmoved Mover differed from the God of Islam, and that *falsafa* omitted such essential information as the Last Judgment, the resurrection of souls, and the dissolution of the world. He allowed the practice of *kalām* where helpful in persuading wavering believers of a rationalistic tendency; but, in general, he thought that theology and the Sufism to which he inclined would do better without it. His position had the strength of logic and of a new Seljuk ruler in Baghdad.

The last stage in Aristotle's journey through Islam jumped two continents, from Persia to Spain, and over a century, from Avicenna to Ibn Bājja (Avempace), the long-serving vizier of the governor of Granada, and Ibn Rushd (Averroes), one-time *cadi*, or religious judge, in Seville and Cordoba. Both insisted on a purer Aristotle than al-Kindī's. Averroes made his career as a physician and protégé of the Almohad Caliph Abd al-Mu'min before becoming *cadi*. His literalist renderings of Aristotle and his opposition to al-Ghazālī's teachings made him the target of traditionalists, who eventually secured his condemnation by a court in Cordoba.

The schism between astronomy and physics appears to have bothered Averroes for most of his philosophical life. As a preliminary to their resolution, he tried to separate Aristotelian cosmology from the Neoplatonic and Islamic accretions it had acquired since the time of al-Kindī. That this mélange had become the default *falsafa* appears from the fable of 'Hai Eb'n Yockdan', as its hero was named in its English translation. Abandoned on a desert island as an infant, Yockdan had plenty of time to think about nature and his place in it. Step by logical step, he invented a Neoplatonic universe on which, after instruction by a passing holy man, he successfully grafted Islam. Averroes declared war on Yockdan. He removed the chain of creating and created beings between the One and the lunar Intelligence, restored the eternal cosmos and the Indifferent Mover, severely criticized both Avicenna and al-Ghazālī, and allowed that Aristotle had possessed as much of the truth as a man can obtain without revelation. One of these truths was that although Ptolemaic astronomy might be good mathematics, it had nothing to do with the real world. The purges of 'The Commentator' (Averroes) thus left the Aristotelian corpus much as the Arabs had found it 400 years earlier. It might be said of the gigantic effort of *falsafa* what Omar Khayyam said of himself after hearing saints and doctors dispute: '[I] evermore came out by the same door as in I went'.

The blunting of *kalām* and discouragement of *falsafa* coincided with an important change in the Islamic educational system. The Baghdad House of Wisdom, with its library and observatory, and its grander version in Cairo built by the Caliph al-Hakīm around 1000, were the high end of a type reproduced in several places in Iraq and Fatamid Egypt. Most were little more than libraries, though some offered ink and the ubiquitous paper to readers. Originally given over to the 'foreign sciences', the libraries either died out or converted to cultivating the 'Islamic sciences' of law, religious studies, and Arabic philology. They did not constitute an educational system.

From the 11th century on, elementary education, which had been the business of mosques, took place largely in charitable institutions controlled by the family of the donor. These mainly Sunni *madrasas* excluded the foreign sciences that had flourished in the largely Shia, Persian-inspired, libraries. The great conquerors in the East, Hūlagū Khan, Timur, and the Seljuks, were great founders of *madrasas*, which helped to inculcate a standard faith and reduce religious controversy. Teaching relied on staggering feats of memorization. It is said of one overachiever that he died having dictated only 30,000 folio pages of the texts he knew by heart.

Mixed mathematics

Astronomers arrived at Averroes' view of Ptolemy's constructions by mastering the translations prepared by denizens of the Baghdad House of Wisdom. Mastery included checking the parameters transmitted in the *Almagest*. After eight centuries, discrepancies between Ptolemy's predictions and observation had become conspicuous. To reduce discrepancies, astronomers working in the Islamic East laboured to improve the parameters and the methods by which Ptolemy had deduced them. Thus humankind came to have good values for the inclination of the ecliptic, the position of the Sun's apogee, the precession of the equinoxes, and other desiderata.

The eccentric astronomer Ibn Yūnus, who served the Fatamids and died faithfully in 1009 on the day he had calculated would be his last, specified one reason for supporting his studies: 'observation of the stars agrees with religious law, for it allows us to know the time of prayers and of the sunrise and sunset that mark the beginning and end of fasting'. Hence Islamic astronomers investigated assiduously topics of only passing interest to Ptolemy, like the duration of dawn and dusk, and conditions for glimpsing the first appearance of the new Moon. Also, astronomy taught the direction to Mecca, the *qibla*, from any place assigned, when to plant, and how to get from one place to another for commerce, pillage, or pilgrimage. Thus, in contrast to *physica*, the study of the stars, gently recommended in the Qur'ān (6.97, 10.5, 16.12, 16, 71.15–16), had the support of religious leaders. It also had the support of secular rulers for its help in imposing religious conformity and in guiding caravans, and for its foundational role in astrology. Ibn Yūnus and his primary patron, the Fatamid Caliph al-Hakīm, were devoted to the subject. Ibn Yūnus's huge *zij*, or handbook of astronomical information, contains tens of thousands of bits of astronomical-astrological data.

Even if they did not outlast their founders, observatories supported frequent determinations of the positions of the luminaries and planets. Whereas Ptolemy had calculated his parameters from very few strategically timed observations, his successors in the East designed observational programmes to last for an entire circuit of Jupiter (twelve years) or, even more optimistically, of Saturn (thirty years). The Islamic lands were fruitful in astronomers, *zijēs*, and monumental instruments, such as al-Khujandī's sunken super sextant, with a radius of 20 metres, located 12 kilometres south of Teheran at Rayy. With this giant, paid for by the local Persian strongman, he made exact determinations of the obliquity of the ecliptic.

By 1050, Islamic-Ptolemaic astronomy had reached its apogee. Astronomers had corrected its parameters via observations with substantial instruments, invented new ways to deduce the parameters from observations, simplified its mathematics with plane and spherical trigonometry, and replaced the cumbersome ancient notation with 'Arabic' (that is, Hindu) numerals. What more was there to do? The approaches of three Persian mathematicians of the mid-11th century—al-Bīrūnī, al-Khayyāmī (Omar Khayyam), and Ibn al-Haytham (Alhazen)—will indicate the options available.

Al-Bīrūnī, whose erudition was extraordinary even for a Muslim scholar, suffered the peripeteias of Islamic savants dependent on courts. Born and educated in Khwārazm, he fled civil war to seek patronage in the fragmenting kingdoms around him. He met al-Khujandī, observed at Rayy, and found temporary employment with fleeting strongmen, one of whom built him an observatory. After the murder of this Maecenas, al-Bīrūnī came under the protection of Sultan Mahmud of Ghazna (Afghanistan), who knew so little about astronomy that he rejected as heresy the assertion that in the far north the Sun sometimes does not shine for days. Al-Bīrūnī enlightened him. Mahmud reciprocated by giving al-Bīrūnī the opportunity to become the sage of the age. Following the Ghaznavid armies to the east, he mastered Sanskrit, determined geographical positions, and brought back much miscellaneous Indian lore.

Most of al-Bīrūnī's writing concerns astronomy, geography, and geodesy. Although Ptolemaic in conception, his astronomy analyses other views, for example, the possibility that the apparent motion of the stars arises from a rotation of Earth. He gave an Indian as well as a Greek source for this idea and reported Ptolemy's objections to it (bodies dropped from a tower would land west of its foot) and the response (all bodies on a rotating Earth would participate in its motion even when falling). As many others would do, al-Bīrūnī accepted Ptolemy's objection.

Khayyam also suffered the vicissitudes of the Islamic savant for whom good fortune, 'like snow upon the desert's dusty face', never lasted long. Born in Khurāsān shortly after the Seljuks conquered the province, he spent his most productive years serving Sultan Malik-Shāh in Isfahan. The sultan and his vizier supported an observatory, where, under Khayyam's direction, a group of astronomers issued their own *zij* and calculated a tropical year closer by three parts in ten million to the truth than the Gregorian rule. The idyll ended with the death of Malik-Shāh and the assassination of his vizier. The new regime beggared the observatory and cracked down on Khayyam's Avicennan world view and poetical free thinking. The line of astronomy he represented, which sought progress in the next place of decimals and required ongoing dependable financing, thus came to an end in Isfahan.

The third example, Alhazen, took the bold position, later pushed by Averroes, that Ptolemaic astronomy had to be reworked to conform to physical principles. He seems to have had unusual confidence, since he proposed to the Fatamid Caliph al-Hakīm—the patron of Ibn Yūnus—a plan to regulate the flow of the Nile. When the plan failed, Alhazen thought it advisable to feign the insanity that his wild project suggested. He recovered his wits when al-Hakīm died and directed them to optics as well as to the shortcomings of Ptolemy. Insisting that the rules of astrophysics required constant rotations of material shells or spheres around their own centres, and that no void could exist in the heavens, Alhazen returned to the nested globes and marbles of Ptolemy's *Hypotheses*.

The three directions of astronomy represented by al-Bīrūnī, Khayyam, and Alhazen came together briefly and dramatically in the work of Nasīr al-Dīn al-Tūsī. From a family of Shia jurists, al-Tūsī received a full Islamic education from several masters, including a follower of Avicenna. Escaping political turmoil in his homeland (like Khayyam he came from Khurāsān), al-Tūsī accepted protection from a tough Shia sect that specialized in murder. In the company of these Assassins and their grand master, 'the Old Man of the Mountains', he wrote many important tracts, some on ethics, which his hosts certainly needed, others on logic, philosophy, and mathematics. After a quarter century in his unusual academic setting, al-Tūsī upgraded to the service of Hūlāgū, who destroyed the Assassins in 1256, conquered Baghdad and terminated the Abbasids in 1258, and established his power from the borders of the Byzantine Empire to the fringes of China.

Though a little rough, Hūlāgū had an interest in astronomy and astrology, and encouraged al-Tūsī to gather up manuscripts before his less learned followers ate them. (Perhaps Hūlāgū owed his civility to the Nestorian Christians, from whom his father had chosen his mother and he his favourite wife.) Hūlāgū erected an observatory for al-Tūsī in the new Ilkhan capital of Marāgha in Azerbaijan. This institution, which Hūlāgū not only paid for (which was to be expected) but also, exceptionally, endowed, attracted several excellent astronomers, boasted a library and a librarian, and housed several large instruments including a mural quadrant and an armillary sphere. One of its first products, twelve years in the making, was, of course, a *zij*. Al-Tūsī and his collaborators disliked the Ptolemaic models on which they had to base their calculations, and, taking their manifesto from the Aristotelian *physica* eclipsed since al-Ghazālī's victory over Avicenna, developed new planetary models. This departure, imitated in the 14th century by Ibn al-Shātir of Damascus, whose day job was tracking religious time, turned out to be a useful step, when, using the same models, Copernicus took a greater stride.

With some imagination, a savant can find verses in the Qur'ān (15.16, 16.20, 24.35) that encourage the study of astrology and optical phenomena. The greatest investigator of light during the Islamic Middle Ages was Alhazen. As in his astronomy, so in his optics but to better effect, he attempted to combine Aristotle's *physica* with geometrical models. Thus, in contrast to most mathematicians, he ascribed vision to rays from a luminous body entering (rather than leaving) the eye. He distinguished between primary light (from self-luminous bodies), secondary light (from all points on a body illuminated by primary light), reflected light, and refracted light. This allowed him to state that the Moon shines by secondary, not reflected, light, and so put a mathematical gloss on Plutarch's description of the lunar surface.

According to Alhazen, rays proceed rectilinearly in all directions from every point on a body shining by primary or secondary light. How then can all the forms and colours they bring to the eye be received without confusion? He answered that rays striking the surface of the crystalline humour perpendicularly dominate the image. As for reflected and refracted light, he confirmed in detail the equality of the angles of incidence (i) and reflection, and gave qualitative rules relating i to the deviation $d = i - r$ suffered in refraction, r being the angle of refraction.

Many mathematicians who wrote on astronomy also wrote on mechanics, notably, in the Baghdad school, al-Khwārizmī, whose name inspired an English common noun ('algorithm'), and Thābit ibn Qurra, a money changer who became a master of all the mathematical sciences; and, among the Eastern astronomers, al-Bīrūnī, Khayyam, and Avicenna. They built on Archimedes' theories of the lever and floating bodies, Aristotle's concepts of weight and motion, and pseudo-Aristotle's *Mechanica*. The interest of their work lies in its comparative realism: they endeavoured, more successfully than in their astronomy, to unite the physical and the mathematical. They did not neglect the weight of the beam when analysing the balance or lever; and they invented or perfected a 'balance of wisdom' with which to measure the specific gravity of objects submersible in a liquid. For two-component alloys, algebraic manipulation of the measurements supplied a more convenient way of detecting counterfeits than Archimedes' method of dropping them into his bathtub.

The most detailed account of Islamic statics and hydrodynamics comes in al-Khāzinī's *Book of the Balance of Wisdom*, which includes an extensive table of specific gravities made by al-Bīrūnī. Al-Khāzinī, who wrote between 1115 and 1130, was the well-educated Byzantine slave of the treasurer of the Seljuk court at Merv in Khurāsān. He debuted in the normal fashion of court astronomers, with a *zij* dedicated to the ruler. He constructed the most precise of balances, inspired by duty to his master the treasurer, who worried about counterfeit gems and alloys, and by pious curiosity about the Day of Judgement, when our deeds will be weighed to the last scruple. Al-Khāzinī's balance employed five weighing pans and reached an accuracy he estimated as one part in 60,000.

Al-Khāzinī's analysis of the stability of balances suspended from axes above, through, or under the beam's centre of gravity makes the weight of a body depend on its position in two distinct ways. It is the greater the further it hangs from the axis of rotation, and the less the higher the specific gravity or density of the medium it occupies. Invoking a strict though inappropriate analogy to the 'weight' (that is, torque) of a body on a balance arm, which runs from zero on the axis to a maximum at the arm's end, al-Khāzinī made a body's weight a maximum at the concave surface of the lunar sphere and zero at Earth's centre. The concept of 'weight according to position' still figured prominently in Aristotelian *physica* when Galileo studied it.

That was around 1589, just when astrological doctrine elaborated under the early Abbasids by the Jew Māshā' Allāh predicted an important alteration in human affairs. His astral history turned on the recondite ideas of grand conjunction and trigon change. A trigon is a group of zodiacal signs 120° apart; a grand conjunction brings together Jupiter and Saturn at twenty-year intervals; after 200 years or so, during which these reunions take place within a given trigon, they move over to another, causing great disruptions (see Figure 6). Do the transitions physically produce disasters or only announce them? Although the Qur'ān (15.81) suggests and reason confirms that the heavens merely advertise, some

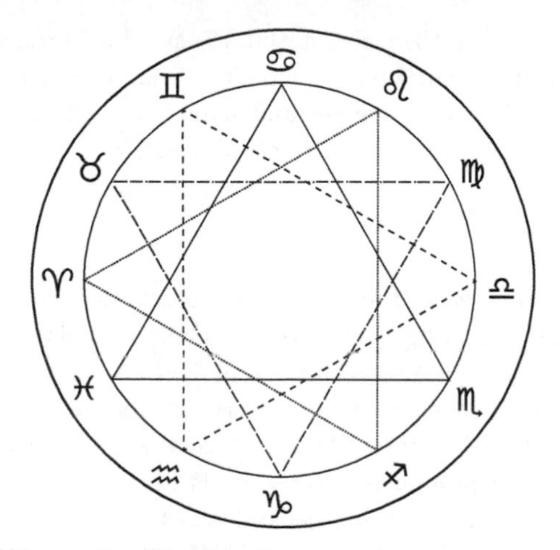

6. Astral history. Grand conjunctions occur every 20 years, trigon changes every 200 years, and cycle changes every 800 years, more or less. The solid triangle connects the watery signs, the dashed the airy, the dotted the earthy, and the dot-dashed the fiery.

astrologers, who identified each trigon with one of the four terrestrial elements, predicted disasters related to the qualities of the element characterizing the receiving trigon. Thus when Jupiter and Saturn first come together in the watery trigon they very probably will cause a flood. The greatest effect of trigon shifts occurs after 800 years when the conjunctions complete a full cycle. Galileo arrived eight centuries after Māshā' Allāh.

Does astrology work through physical agency and therefore belong to *physica*, or only give notice of future happenings through signs predictable by astronomy, and thus fall to mathematics? In this guise the old problem of the relations between mathematics and physics acquired the complication of whether astrology's predictions, if correctly interpreted, could be avoided. The Qur'ān, which declares frequently that only God knows the future, hints at a possibility of avoidance (16.12, 27.65, 67.5–9). Avicenna and Averroes appealed to these declarations to convict astrology of futility. But the early adepts, like Māshā' Allāh, al-Kindī, and Thābit ibn Qurra, who worked the literal star lore of Persia into Islamic astrology, accepted the fatalistic principles and predictions of their art. The strictest advocate was the leading Muslim astrologer Abū Ma'shar, another Persian from Khurāsān, who came to Baghdad close to the beginning of the reign of al-Mamūn.

Further illumination on the question of celestial influence can be found in medical practice. Astrologers associated a zodiacal sign with body parts, thus Aries with the head, Taurus with the neck, Gemini with the shoulders, down to Pisces and the feet. As anyone can work out by observation of the tides and the menstrual cycle, the Moon has some influence over fluids. Hence it might cause unwanted interference during medically sanctioned bleeding. To minimize the risk, physicians avoided bleeding from any part of the body associated with the zodiacal sign through which the Moon was passing. This abstention is more plausibly construed as a response to a risk from a real physical force than to a notice delivered by abstract signs. Hence medical practice exposed another reason to integrate terrestrial physics with the mixed mathematics of the heavens.

Departures

In glaring contrast to the philosophers of Greece, members of the Baghdad House of Wisdom from al-Kindī on, and polymaths like al-Bīrūnī, Alhazen, and Avicenna, wrote on technological and engineering subjects. The Islamic states undertook large-scale works—for example, the construction of Baghdad, with its mosques, gardens, and palaces—and large-scale wars, with their demands for weapons and fortifications. The decorations of the buildings and their furnishings incorporated geometrical designs of great beauty and complexity, and the strength of the weapons relied on unique and intricate metallurgical processes. The famous swords made of 'Damascene steel' combined beauty with utility. The high-end metal industry nourished by military requirements also supplied parts for advanced engineering projects and scientific instruments. Of these latter, the planispheric astrolabe may stand as the emblem or logo of Islamic science.

An astrolabe consists of several delicately inscribed brass plates contained in a stubby cylinder. It could be adjusted to show in a plane the places of the Sun and stars as observed in the stellar sphere. The operator needed only to turn the open-work 'rete' until one of its star points, or the Sun's position on the ecliptic, came above the appropriate circle of altitude or azimuth on the lower plate or 'tympan' (see Figure 7). A good astrolabe carries several tympans engraved on both sides for use at different latitudes. The concept of the instrument is Greek; its realization in brass, Islamic. The oldest extant example, from Isfahan, dates from the 10th century.

Once set, the astrolabe gives the time of day, the time and direction of sunrise and sunset, and, with additional information, the *qibla*. It can also serve as a compass for navigation. Thus it elegantly met the needs of the religion founded by the merchant Muhammad: it gave the times and orientation of daily prayers, guided pilgrims' steps, and helped caravans cross the desert. In addition, the sighting bar rotatable against the circular scales on the back of the instrument can measure the radius of Earth. Al-Bīrūnī took his astrolabe to the top of a mountain in India adjacent to a plain and measured the angle θ between the vertical and the horizon. Any neophyte in Muslim trigonometry could then work out r from $\sin \theta = r/(r + h)$, h being the mountain height, which al-Bīrūnī measured by a standard surveying technique (see Figure 8). His result confirmed the value obtained by al-Mamūn's mathematicians, although 'their instruments were more precise, and their labour to obtain it of an extremely exacting and fastidious nature'.

We return briefly to the House of Wisdom, where the Banū Mūsā—three sons of a bandit named Mūsā ibn Shākir, who died

7. Computer in brass. An Iranian astrolabe showing the rotatable rete with its star points and, underneath it, a tympan with the fixed reference circles of the observer.

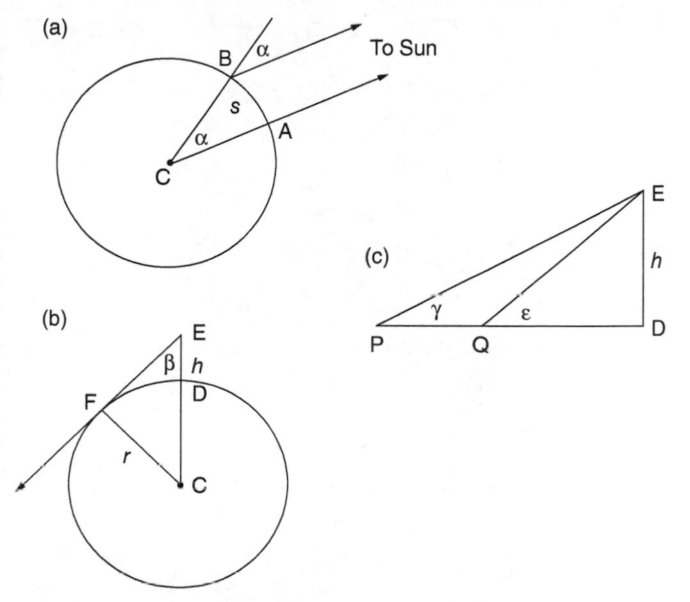

8. Size of the Earth. (a) Eratosthenes' method: A is Aswan, B Alexandria, α the measurable zenith distance of the Sun at B, *s* the separation of A and B as determined by walking, sailing, or guessing. (b) Al-Birūnī's method: E is the mountaintop, EF the horizon through F, *h* the mountain's height; the rest is trigonometry. (c) *h* determined from measurements of the angles γ, ϵ from the stations P and Q and the paced-out distance between them.

an astrologer at al-Mamūn's court—diverted themselves from mathematics to improve the devices of Hero of Alexandria. They added siphons that, among other feats, appeared to discharge wine and water separately after imbibing them mixed. The greatest mechanician in this tradition worked for a small Turkish dynasty that ruled land around the upper Tigris. He was al-Jazarī, whose *Compendium of the Theory and Practice of the Practical Arts*, 'the most important document on machines from ancient times to the renaissance from any cultural area', describes feedback control and other novel techniques.

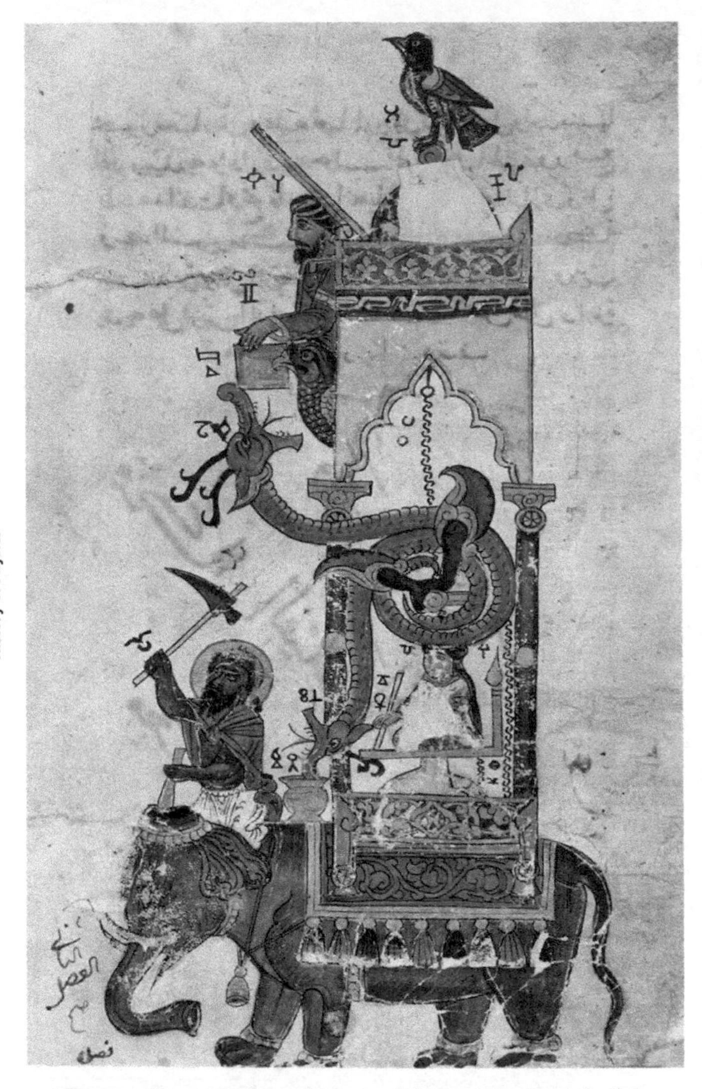

9. The magnificent elephant clock. An elaborate automaton built by al-Jazarī that celebrates every half hour, day and night.

A fine example of his automatons, the elephant clock, carries an open domed tower with a phoenix on top (see Figure 9). Half way down the tower under each falcon a dragon rides on a horizontal axis. The elephant's body contains a large vessel full of water in which a perforated wooden bucket, loosely attached by ropes to some gearing, floats. Every thirty minutes, the leaky bucket sinks, pulling the ropes and engaging the gears, which make the phoenix whirl, causing a ball to fall through a falcon head into a dragon's mouth. The lucky dragon swallows the ball, swings on its axis, rewinds the ropes, and brings the bucket back to the surface. The clock continues to play every half hour until it has exhausted its balls.

Al-Bīrūnī mentions that gears were applied to astrolabes. The earliest surviving example, which is also the earliest extant intact gear train, dates from 1221. Five finely cut gears mounted inside the backplate drive images of the Moon to indicate its place and age. Inclusion of information about the Moon's zodiacal position made the astrolabe a medical instrument, since it allowed the doctor to discover with a twist whether or not to cut the patient. If the twist were provided by a falling weight controlled by an escapement, the geared astrolabe would be a mechanical clock. Such clocks, and many subsequent imitations, still exist, inadvertently advertising, via their astrolabic faces and solar and lunar pointers, the logo and achievements of medieval Muslim science.

Chapter 3
Domestication in Europe

The domestication of Greek and Arabic *physica* and mixed mathematics in the Latin West took some 400 years from the first translations in the 12th century until well into the 16th, when Archimedes and Ptolemy were printed and Aristotle's ancient rivals revitalized. With the generation of Galileo, Kepler, and Francis Bacon, *physica*'s place in the body of knowledge began to slip, although the Aristotelian world picture still hung securely, if awry, in universities and theological seminaries. The slippage owed much to social factors associated, as in Islamic times, with the needs of newly centralizing states, and with the discovery of new worlds on the Earth and in the heavens.

Stirrings

While Alhazen wrestled with geometrical optics, two polymaths in the West tried to prove that the sum of the angles in a triangle is a straight angle. Not knowing Euclid, they had neither the proof nor the concept of proof, and ended by lining up the angles cut from triangles drawn on parchment. Although they represented the mathematical knowledge of the learned of their time and place, a few savants existed then and there who could handle hard computations easily. They were the calculators of times of prayer, dates of movable feasts, and calendrical coordination. Their master, the Venerable Bede, a Northumbrian monk who died in

735, wrote on *physica* as well as *computus*, and dated events *anno domini* from the year now taken as the first in the Christian era.

Bede's *physica*—his understanding of the motions of the planets and the luminaries, of the rotations of the heavens, and the interconvertibility of the elements—came from Pliny's *Natural History* and the commentaries on the *Hexaemeron* by Saints Basil and Ambrose. From the same sources Bede learned that 'the ocean tide follows the moon as if the luminary drew the waters behind it and then pushed them back' (thus the saints) and that, owing to the time required for celestial events to affect Earth, a delay occurs between high tide and Moon's meridianal passage (Pliny). Local knowledge then prompted Bede's important deduction, once acclaimed as the only original contribution to *physica* from the Latin West in 800 years: the difference in time between the meridian passage of the moon and the succeeding high tide is a constant at any place. The riddle of the tides thus acquired an additional complication.

Bede's knowledge made its way to the Continent via Alcuin of York, who served as Charlemagne's chief advisor in civilization. Alcuin revived the curriculum of the liberal arts, and stimulated an interest in recreational mathematics. A booklet attributed to him comprises fifty-three perennial problems, for example, river crossings like the familiar puzzle of the missionaries and the cannibals, and divisions of property. His arithmetical games may have inspired some students to go on to *computus*, on which he wrote a tract; but when the 'Carolingian Renaissance' petered out, it left its polymaths ignorant of geometry.

One of Alcuin's tasks was to write against a Christian heresy prospering in Umayyad Toledo. Umayyad culture had beguiled these heretics with 'the Arabic pomposity of language', with gardens, libraries, and palaces, with scholars, artists, and musicians, and, for the technically minded, with improvements in metalworking, agriculture, waterworks, and trade. It was not Alcuin who saved these straying sheep, however, but small Christian Spanish kingdoms that profited from the usual divisions within Islamic states to capture Toledo in 1085. The Christian scholars who followed the Christian generals did not care for Arabic poetry. They wanted to learn the *physica*, philosophy, astronomy, and mathematics they knew they lacked; as one of them acknowledged, 'our civilization is infantile in comparison with theirs'.

The earliest large-scale translation work in Spain mirrored the effort in Baghdad 400 years earlier. The archbishops of Toledo played the role of the early Abbasid patrons. Translators again invented technical terms via transliteration, bequeathing to English zenith, nadir, azimuth, almanac, algebra, Aldebaran and many other star names, alkali, alembic, elixir, alcohol, things to eat (apricot, artichoke, sherbet), and agents of Islamic prosperity (bazar, tariff, admiral, arsenal, cotton, muslin, damask). And, as the scholars of Baghdad were recruited throughout the Islamic world, so Westerners wanting wisdom came from every corner of Christendom. Besides the Greek corpus, the scientific authorities awaiting their study included, to mention only names already encountered, Albumasar (Abū Ma'shar), Alhazen, Avicenna, Averroes, Avempace, Alfarabus, Alkindus, and Māshā' Allāh.

The first Latins who tried to apply Islamic astrological physics deduced from an upcoming trigon change accompanied by a congregation of planets in Libra (an airy sign) that great winds from the East would dump deserts of sand on Europe. The neophytes feared the worst. A friendly Spanish Muslim astrologer wrote the Bishop of Toledo a letter of reassurance. The ignoramuses who made the predictions, 'knowing nothing of the virtues of the heavenly bodies and the effects of the five planets and the two luminaries', had reasoned childishly.

By coincidence, the West had just invented an institution fit to master the new learning. This was the *universitas* or guild organized by teachers accredited for their learning, chartered by the Pope, and supervised by the local bishop. In a few cases, Bologna being the most prominent, students formed the corporation and hired the professors. (See Figure 10.) Founded by modern consensus in 1088, Bologna dealt primarily in law; the *studium generale*, comprising at its fullest a preparatory arts faculty and professional schools in theology, medicine, and law, represented by the universities of Paris and Oxford, dates from around 1200. To add further dynamism and coincidence, the energetic new mendicant orders, the Dominicans and Franciscans, decided soon after their foundations in 1216 and 1223, respectively, to establish chairs in the arts and theological faculties. Many of the greatest interpreters of Aristotelian *physica* were members of these vigorous orders. Thus, in significant contrast to Islam, the Roman Catholic Church embraced rationalistic *kalām* and produced a scholastic philosophy in which *physica* had an important place.

Alma mater

Most of the teachers in arts were men engaged in the lengthy pursuit of a doctorate in theology. Their bachelor's ('beginner's') degree conferred the right to teach arts anywhere and thus to produce a new generation of artists, in a sort of academic apostolic succession. Like their Islamic counterparts, Latin scholars had a common language and religion; unlike them, the Westerners had universities that gave institutional continuity and ordered competition in the study of general subjects deemed useful for theology and the lay professions of law and medicine.

Episcopal supervision, clerical professors, and a universal teaching credential did not create uniformity of curriculum or opinion. The arts masters favoured Aristotle's philosophy as the best available knowledge attainable by reason alone. Theologians initially opposed it for its novelty, and also, more soundly, because it taught many things obnoxious to the Catholic faith. And so statutes of the University of Paris enacted in 1215 required candidates for the bachelor's degree in arts to have studied

10. **Stoned students. A scene from the arts faculty at the University of Bologna in the thirteenth century, as recorded on a tombstone.**

Aristotle's logic and ethics, but prohibited the public reading of his *physica* and metaphysics. After the *studium generale* in Toulouse (founded in 1229) advertised that it allowed these works, the Pope, Gregory IX, ordered the University of Paris to separate the useful material in them from the erroneous and scandalous.

Soon the arts masters at Paris were teaching an Averroist Aristotle and defending their presentations of erroneous and scandalous doctrines as fulfilling a commentator's duty to the author. This historical method of introducing naturalistic *physica* continued through the time of Galileo's teachers. However, it did not prevail in medieval Paris. In 1277, the Bishop, Etienne Tempier, having received the recommendations of a committee of theologians, prohibited over 200 harmful propositions they claimed to have found in Aristotle. With this apparent abrogation of academic freedom, Tempier liberated the Paris faculty from a slavish literal interpretation of the texts and opened the way to wider and wilder speculation. The University of Oxford, which, together with Paris, produced the most advanced *physica* of the later Middle Ages, adopted proscriptions similar to Tempier's.

The enduring correction of Aristotelian philosophy through revealed truth was the work of a cosmopolitan Saint, Thomas Aquinas, born in Italy, educated in Paris and Cologne under another saintly exponent of Aristotle, Albertus Magnus, and active as a Dominican master in Paris and Rome. To adjust Aristotle to Revelation, Thomas replaced the unit Universal Mover by the triune Christian God and the Intelligences by angels, much as the Neoplatonists had done, and eliminated the naturalistic errors about Creation, the soul, the vacuum, the displacement of the world, and so on. He retained the quintessential celestial spheres, the four-ring circus of terrestrial elements, the meteorology, and all the paraphernalia of form and matter.

Dante took Thomas's world picture as the great theatre for his journey through Earth to the Devil's umbilicus and thence through the antipodes, up Mount Purgatory, and past the planets to the divine presence beyond the sphere of the stars. On the way, Dante stopped at the sphere of the Sun for a talk with the shade of St Thomas and his circle of eleven sages. Bede was among them, furthest from Thomas; the nearest shades belonged to his teacher Albert and to Siger de Brabant, a champion of Averroist Aristotelianism and Thomas's unrelenting opponent in life. Their juxtaposition in heaven suggests that the claims of reason deserve a hearing even when in apparent conflict with revealed truth.

Scholastic theologians expounded *physica* to clarify doctrinal problems. Take charity. What prompts a philanthropist to give? Since in Aristotelian philosophy the question fell under the category of motion, theologians of charity could exploit concepts developed to analyse physical change. If a shove from the Holy Spirit causes the philanthropist to act, fundraisers should expect giving to slacken just as a struck bell falls silent. Still, the ringing persists briefly and sometimes so does charity. What causes this persistence? The clapper transfers motion to the bell as the Holy Spirit conveys a charitable impulse to the philanthropist. Thus the concept of conferred impetus, whether fleeting or enduring, settled into medieval *physica*. Applied to locomotion, it allowed arrows to fly without being pushed by the air and explained the acceleration of falling bodies as the sum of the impulses acquired in each minute or ell of descent.

The outlawing of Averroistic Aristotelianism encouraged criticism of the fundamental concepts of form and essence. Does 'dogginess' exist or is 'dog' just a convenient name for a group of animals? The second alternative, nominalism, had its major champion in William of Ockham. In lectures he gave at Oxford as an inceptor (a bachelor of arts working at theology), he insisted that essences or common natures do not exist, and that general propositions about groups or classes cannot be known to be true. The 'Venerable Inceptor' (so called because he never finished his degree) held that nothing constrains God's action but the impossibility of compassing a contradiction; that only experience, therefore, and not a priori deduction, can establish what exists; and that in explaining the apparent relations among things the fewest possible causes should be invoked ('Ockham's razor').

Nominalism favours description over explanation. Hence we find in 14th-century Oxford an innovation, introduced by Ockham's contemporary, Thomas Bradwardine, for discussing the grand question, how the form of a substance changes? Rather than vex themselves over the causes of change, Bradwardine and others at Oxford's Merton College distinguished between motions at constant speed and at uniform acceleration, and deduced that in locomotion the total distance traversed from rest in a given time at constant acceleration is half the distance that would be traversed at the maximum speed in the same time (the 'Mertonian rule'). They made no direct applications of their rule, not even to free fall.

The most distinguished teacher of *physica* at the University of Paris in the 14th century was Jean Buridan, now known primarily for his ass—a donkey with a mind so logical and rigorous that it starved to death for lack of a sufficient reason to choose between two equally attractive bales of hay. Buridan was a nominalist concerned to carve a space for a living Aristotelian *physica* between the stultifications of Averroism and Ockham's voluntarism. In a bold step he applied the concept of impetus to the celestial spheres. They would not need angels or Intelligences to inspire them if God gave them an initial push and the quintessence conserved the impetus. Buridan observed further that the Earth must jiggle owing to erosion and the self-movement of living creatures, which constantly shift its centre of gravity. Perhaps the Earth might even revolve around the centre of the universe and allow the heavens to rest? These notions should be appreciated as indications of the sorts of things that free thinking within an Aristotelian framework could produce.

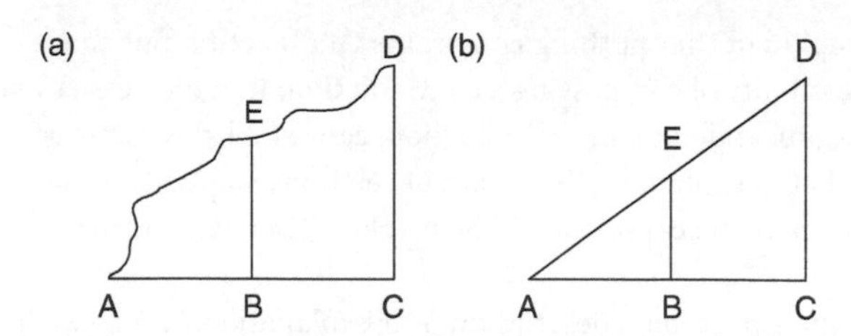

11. Motion pictured. (a) AB the 'longitude,' usually time or space, BE the corresponding 'latitude,' or intensity of a changing form; ABCDE, the 'total change' of the form over the entire longitude AC. (b) The case of motion at constant acceleration.

Buridan's prize student, Nicole Oresme, invented a convenient and influential, though misleading, representation of the Mertonian rule. In it a horizontal axis or 'longitude' represents space or time, its associated vertical axis or 'latitude' the corresponding degree of an accidental quality like place, heat, or colour (see Figure 11). In locomotion, the accident was place; the latitude, velocity; and the longitude, indifferently space or time. The 'latitude-of-forms' picture of motion at constant velocity is a rectangle, at constant acceleration a triangle or trapezoid.

One of Aristotle's arguments against the possibility of a void makes velocity v proportional to the mover's push P and inversely proportional to the resistance R of the medium, from which the absurdities follow that velocity would be infinite in a vacuum and finite even if $R > P$. Avempace had evaded these consequences by supposing $v \propto (P - R)$. Bradwardine and Oresme considered an alternative often rendered anachronistically as $v \propto \log(P/R)$. It had the theoretical merit of preserving Aristotle's argument against the void and making $v = 0$ when $P = R$.

The teaching that, according to later judges, made the 14th century the 'Age of Buridan' occurred entirely in the Arts Faculty of Paris. Buridan did not go on to theology but stayed in arts. In a development of capital importance for all liberal studies, arts faculties strengthened against professional ones. Long-term cultivation of special subjects became possible and allowed advancement with age, reputation, and administrative responsibilities. Whereas Bradwardine and Oresme pursued clerical careers (Bradwardine served for a month as Archbishop of Canterbury, Oresme a few years as Bishop of Lisieux), Buridan rose to Rector of the Arts Faculty and received benefices to support his studies. But death levels distinctions, and the Black Plague indiscriminately ended the lives of the Rector, the Bishop, the Archbishop, and the Venerable Inceptor.

Mixed mathematics

Like the impetus theory of charity, statics (*scientia de ponderibus*) and optics (*perspectiva*) had a bearing on Christian religion. Medieval statics solved the pressing problem of *gravitas in situm*, the tendency T of a body to slide down an inclined plane, which related to arches and domes. Thābit ibn Qurra and Avempace had made T equal to the weight of a body W diminished by the ratio of the plane's height h to its length L, $T = (h/L)W$. Their Western follower, Jordanus Nemorarius, who wrote in the early 13th century, combined the dynamic approach of the pseudo-Aristotelian *Mechanica* with a geometrical analysis in the style of Archimedes; his packaging of ancient and Arabic contributions to the science of weights served the West for two or three centuries. It was said that experts in the *scientia de ponderibus* could weigh divine grace if they implemented Jordanus's theory faithfully.

The light of Creation, the Rainbow of the Covenant, and the filtered colours within Gothic cathedrals inspired Christians to the study of optics. Robert Grosseteste, trained in Oxford and Paris, acted as director of studies in *physica* and mathematics to the Oxford Franciscans. Beginning at the beginning, Grosseteste imagined the light of Creation to have spread itself like a Neoplatonic emanation, creating space, the celestial spheres, and the rest of the cosmos. This *lux* (created light) manifested itself depending on where its *lumen* (propagating light) operates. Grosseteste's *lux* found its ideal medium in Roger Bacon, also educated in Oxford and in Paris, where he taught Aristotle's *physica* and, in 1257, became a Franciscan. This career change proved to be a major error. The General of the Order, who became Saint Bonaventure, did not support studies not bearing directly on theology, and rejected two branches of *physica*, alchemy and astrology, that Bacon esteemed. His superiors jailed him in 1277, the year of Tempier's prohibitions.

Bacon's development of Grosseteste's optics invoked the 'multiplication of species', the spherical propagation of *lumen* from points of *lux*, which served also as a physical model of astrological influence. Bacon's *Perspectiva* incorporated Alhazen's optical principles, including image formation, except for intromission. Attempting then to multiply his influence, Bacon designed a project to refute unbelievers. It required knowledge of *physica* and of a *scientia experimentalis*, 'wholly unknown to the general run of students'. This novel science, which Bacon did not practice much himself, would investigate such means for curing heretics as alchemical transmutations, everlasting lamps, and irresistible explosives.

Bacon's observations included the puzzling finding that the maximum height of the rainbow is 42°. In the first ever account of rainbow physics, God had written, 'It shall come to pass, when I bring a cloud over the earth, that the bow shall be seen in the cloud.... And the bow shall be in the cloud.' Subtle is the Lord! Did He mean that the bow exists in the cloud or that we only see it there? Bacon chose the second option and deduced that the observer creates the bow from rays undergoing reflection and refraction in rain droplets. Variations of this idea occur in influential works by other theologians: in a *Perspectiva* derived from Alhazen by Erazm Ciołek Witelo (a Polish theologian educated at Paris and Padua), and in a treatise *De Iride* by Theodoric of Freiberg (a Dominican who studied and taught in Paris), which traced rainbow rays reverently and geometrically from the Sun through their reflections and refractions to the eye. The 42° remained a mystery.

Fresh imports

As the new universities assimilated their Greek-Muslim inheritance, other important material arrived with people fleeing a Byzantium wrecked by crusaders and threatened by other barbarians, and with investigation of the archives of Italian churches and monasteries. Many Greek manuscripts thus emerged in Italy before the Eastern Roman Empire finally fell to the Turks in 1453. They included the works of Plato and the Neoplatonists, Greek versions of Ptolemy and Archimedes, and excerpts from Stoics, Sceptics, and Atomists. Just as a new institution, the university, coincidentally arose to receive the Arabic heritage, so a new social movement, Italian humanism, which started among learned secretaries serving the princes of new city states, identified itself with Greek language and literature.

The Greek *Almagest* that the most famous of the scholarly émigrés, Johannes Bessarion, later a Roman Catholic cardinal, carried to Italy, brought down the curtain on ancient astronomy. Like timid students who tend to blame themselves for not understanding their teachers, the astronomers of the West supposed that Ptolemy had all the answers. But when Bessarion's protégés Georg Peurbach and Peurbach's student Johannes Regiomontanus (University of Vienna) scrutinized the Greek *Almagest*, they discovered that it did not clear up the problems of astronomy any better than versions made from the Arabic.

While the purified Ptolemy was dashing the hopes of humanist astronomers, other translators were recovering other disappointing truths. Impressed by intellectual émigrés from Byzantium, Cosimo I de' Medici desired to resurrect the Platonic Academy in Florence. Its great project was the translation of Plato, Plotinus, and other Neoplatonic philosophers. Cosimo confided the task to Marsilio Ficino, whom he ordered to begin with the hodgepodge of Christian, Neoplatonic, and Gnostic thought put together in the 3rd century CE under the name of Hermes Trismegistus. Most humanists regarded these writings as very ancient and Ficino, impressed by their apparent anticipation of Scripture, thought that Hermes was contemporary with Moses or, perhaps, Moses himself. Hermes taught among other things that celestial influences can be captured in talismans. A *physicus* willing to experiment enough might master astrological powers.

The Platonic Academy's Latinized Plato spread the Pythagorean doctrine that made mathematics the queen of *physica* rather than at best a handmaiden. In 1464, when Ficino had just started his translations, Regiomontanus promoted the Pythagorean vision in a lecture in Padua on Islamic astronomy and astrology. The lecture extols astrology as the highest, noblest, and hardest branch of mathematics, through which, 'no less than through the other arts, we are set apart from wild beasts'. Regiomontanus inferred from his ability to master the 'angelic art' of astrology that mankind, Hermes' *magnum miraculum*, might stand as high on the scale of being as demons.

Among the astronomers encouraged by this insight was Nicholas Copernicus, born the year Regiomontanus published Peurbach's epitome of Ptolemaic planetary theory (1473). At 23, after studies at the University of Cracow, he went to Italy to learn law and medicine in preparation for an ecclesiastical career. He also learned enough of the humanists' language to translate some Greek poetry. His subsequent position as a cathedral canon in a region contested by Polish and German princes, and also by Prutenic Knights, left him little time to polish his version of proper astronomy. This turned on the objection pushed by Averroes et al.: Ptolemy's use of improper devices like the equant point. Copernicus removed it using constructions that convert circular to rectilinear motion probably taken from Arabic astronomers ('al-Tūsī's couple'), although no persuasive path of transmission has been established. But the boldness and confidence that prompted Copernicus to go beyond removing the supposed blemishes of the equant to place the Sun at the world's centre did not have a counterpart in Islamic astronomy.

The description of the apparent motions of the Sun, Moon, and stars is equivalent whether the Sun or the Earth stands still. By making the Sun the centre of motion, however, Copernicus could do what Ptolemy could not: calculate the distance from the Sun to a planet as a multiple of the radius of the Earth's orbit. His ordering of the planets and his consequent explanations of the limited elongations of Mercury and Venus, and the retrogradations of Mars, Jupiter, and Saturn, were the anchors of his claim to have made a system out of Ptolemy's bric-a-brac (see Figure 12). Copernicus's masterpiece, *De Revolutionibus Orbium Coelestium*, appeared in 1543 with an unsigned preface by a Lutheran controversialist, Andreas Osiander, who saw the manuscript through the press. Anticipating objections from theologians, he rehearsed the usual arguments against taking mathematical constructions as truths of nature. The book did not excite theologians for seventy years. However, it caused an immediate ruckus among *physici*, who saw that it disagreed more profoundly with *physica* than Ptolemy's fictions.

Copernicus addressed some of the obvious objections, like the removal of the Earth from the universal centre of heavy bodies, the lack of a power to move it, the vertical drop of falling bodies, and so on. Some of his answers were physical (separated bits of planets fall to their centre, objects on a rotating planet participate in its motion) and some rhetorical invocations of ancient writers prized by humanists (Plutarch, Pythagoreans, Plato, Virgil, Hermes, Sophocles) who said things favourable to belief in a Sun-centred universe. Copernicus's answers left much to do before an adequate replacement for standard *physica*, with a suitable role for mathematics, could even be sketched.

While the cause of mathematics as natural philosophy advanced with the help of Plato's rediscovered eloquence, humanists

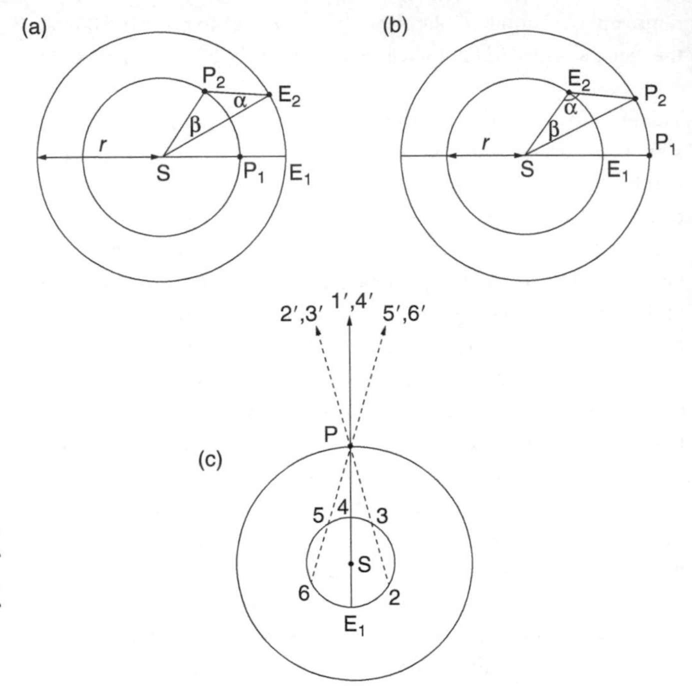

12. Heliocentric advantages. Copernicus allows the computation of the solar distances of the planets P in terms of the Sun–Earth distance *r*. (a) The limited elongations of the 'inferior planets' Mercury and Venus are a consequence of Earth's orbit encircling theirs. (b) The 'superior planets' Mars, Jupiter, and Saturn can be seen anywhere because their orbits encircle Earth's; in both cases, α can be measured and β calculated from the known sidereal periods. (c) The superior planets appear to move backward, or 'retrograde,' against their usual cc motion as the Earth overtakes them, as between positions 2 and 4. For convenience, P is pictured as stationary.

brought forth a threat to Aristotelian *physica* in the form of Latin poetry. This was the long-lost work of Lucretius, found in the early 15th century and printed for the first time in 1473. The notion that our world is but one of an infinite number composed for no purpose by uncreated atoms was not wholesome doctrine for Christians. St Jerome had offered the monkish explanation of the poem's perversity that aphrodisiacs had driven its author insane. Nonetheless, the beauty of Lucretius' descriptions of the couplings of the atoms recommended his poem to all lovers of Latin, and helped to insinuate atomism as an alternative to the *physica* of the schools, the mathematizing of the Platonists, and the Hermetic account of occult influences. Like Averroism, atomism became a dangerous subject during the 16th century, when the various Christian sects readily hurled charges of heresy at one another and Aristotelian *physica* became officially entangled with theology and the dangerous doctrine of the Eucharist.

The tolerance of ambiguity in medieval thought, represented by compromises between reason and revelation, began to give way to more exact forms of knowledge and action during the 15th and 16th centuries. The principal causes of this hardening were the rise of powerful competing orders within the Roman Catholic Church; the Protestant reformation and the Catholic response; the discovery of the new world and its exploration; and the invention of printing. Religious controversy required clear definition of the points at issue among Catholic orders and Protestant sects. The Catholic Church replied by strengthening dogma and discipline through the decrees of the Council of Trent, and by setting up a regime of thought police: the Holy Roman Inquisition (the Holy Office), the Congregation of the Index of Prohibited Books, and the Society of Jesus (the Jesuits).

On the secular side, the European discovery of America intensified the bureaucratic needs of the states competing in the new world. Managing the fleets, keeping track of precious metals, surveying the new possessions, and teaching and improving the arts of navigation opened new opportunities for applied mathematicians already in demand as architects, urban planners, and hydraulic engineers. Calculations depend on hard edges. Large collections of unfamiliar material brought back from exploratory voyages, and reports of strange peoples who did not have the happiness of being European, increased the demand for minds able and energetic in denying and distinguishing. And, to complete this hasty inventory, printing, by replacing costly and imprecise manuscript copying with more or less exact reproductions, gave scholars access to standard texts. Standardization and definition did not resolve disputes, but made it more likely that the parties to them argued about the same things.

Many consequences for *physica* of hardening and standardization may be discovered in Galileo's work. When professor of mathematics at the University of Padua, from 1592 to 1610, he took in private students of aristocratic rank to teach them the use of surveying and measuring instruments. His expertise in designing practical devices enabled him to transform a spyglass of little magnification into a 30-power telescope in a few months. The printing press helped 'the new Columbus' to spread news of his implausible telescopic discoveries—mountains on the Moon, satellites around Jupiter, stars in the Milky Way—as it had given him access to the books of Euclid, Archimedes, Ptolemy, and Copernicus from which he drew inspiration. When he interpreted his discoveries in ways that conflicted with Scripture, the hard-edged institutions of the Catholic Counter-Reformation came into play. Galileo's conclusions catapulted astronomy into the contested arena of *physica* and himself from a professorship of mathematics at the tolerant Venetian University of Padua to the hazardous height of 'Philosopher and Mathematician to the Grand Duke of Tuscany' in the bigoted Medici court in Florence.

Galileo's friend Pope Urban VIII was a cultured Florentine and man of the world. Yet he believed on grounds stronger than Plato's that the unaided human mind cannot arrive at secure truths about the natural world. Urban had followed the logical track of God's Omnipotence and Freedom to the conclusion that, however well our theories fit the facts, God could have contrived to bring about the same phenomena in countless other ways. Galileo countered that his business was not to imagine how God might have done things, but to discover the method He had chosen. Galileo's spirited defence of Copernican theory, made in his *Dialogue on the Two Chief World Systems* (1632) sixteen years after the Holy Office had found heliocentrism contrary to Scripture and 'philosophically absurd', precipitated his trial, condemnation, humiliation, and perpetual house arrest.

What made Galileo a hero to later physicists, however, was not his passive suffering on behalf of Copernicus but his active attack on *physica* under the banner of mathematics. When a young professor at the University of Pisa, he replaced the supposedly Aristotelian formula, $v \propto P/R$, with Avempace's rule, interpreting $P - R$ in Archimedean terms as the difference between the specific gravities of the falling body and the resisting medium. If Galileo dropped weights of different materials from the Leaning Tower as the legend started by his last pupil, Vincenzio Viviani, has it, he would not have expected them to descend at the same speed.

Galileo came to greater enlightenment by abandoning the search for causes. He gave up on Archimedes as he had on Aristotle and entered into a free fall with no goal in sight other than a mathematical description of the journey. He found and confirmed by some mixture of experiment and conjecture that (as he first expressed it in 1604) bodies descending freely from rest cover, in successive intervals of time, spaces proportional to the odd numbers beginning with one. Representing his rule by a graph identical to the medieval picture of the latitude of forms, he had, as he wrote it, $\Delta ABE{:}\Delta ACD = AB^2{:}AC^2$ (Figure 11b, rotated 90°). Since he intended the graph as a representation of mathematical relationships, he had to decide what the latitudes and the triangles stood for. Initially he fell for the obvious choice, taking the vertical axis to be the distance of fall and the horizontal instantaneous velocity. Eventually he realized that the choice led to impossibilities. Taking then the vertical axis as time, he had a successful plot, at the major expense, however, of the counter-intuitive identification of a line with time and an area with distance. With this sacrifice or advance he achieved the now celebrated, but then not obviously useful, result that distance traversed under constant acceleration from rest is proportional to the square of the elapsed time.

Galileo's kinematical method, ignoring forces and causes, was perfectly adapted to the contest over world systems. The circles and epicycles of Ptolemy and Copernicus, which turned without physical force, are kinematical devices. As everyone knowledgeable knew, without a physics (or a theology!) to determine the world's centre, the circles might be referred to Sun or Moon rather than Earth. Hence neither of the chief world systems, nor their chief competitor—the system worked out by the most exact of observers, Tycho Brahe, in which planets go around Sun while Sun circles Earth—could be affirmed to be true on the basis of astronomical observations alone.

Galileo tried to go further, and to reduce the effect of Earth's motion to purely kinematical and geometrical relations, and so defeat the standard physical arguments against it. He devised a kinematical argument to show that only if Earth moved as Copernicus said it did would there be tides, and published his theory as the provocative conclusion to his *Dialogue*. To counter the impression that he believed in the condemned Copernican system, Galileo saved the day but not himself by having the dunce of the *Dialogue* undercut all world systems with Urban's epistemology. The Pope never forgave Galileo for trashing the powerful doctrine with which he had hoped to answer all challenges to Scripture and papal dicta that *physica* could mount.

Galileo situated his *Dialogue* in a place fit for a discussion of liberal arts—a palace on the Grand Canal in Venice. He chose the Venetian Arsenal, a shipbuilding complex, for his *Discourses on Two New Sciences* (1638), which treated the strength of beams and exterior ballistics. Although of little practical value, his geometrical analyses indicated what a quantitative applied physics might look like; and his technique of compounding a constant

13. Applied Physics around 1550. A series of pumps removing water from a mine. The series overcame nature's limit of a 30-foot rise for a single pump. From Georg Agricola's *De re metallica* (1556), a coffee-table book describing rough technology in polite Latin.

velocity imparted by the cannon with an accelerated drop caused by gravity, from which he derived the ideal parabolic trajectory of projectiles, became fundamental to analytical mechanics.

In moving his discussants from palace to workshop, Galileo made common cause with a few university-trained men who took a sustained interest in the physical principles of practical devices. One of them, Georg Agricola, a doctor who practiced medicine on miners, illustrated their techniques so plainly as almost to thrust the concepts of force, momentum, and pressure into the reader's mind (see Figure 13). Another doctor, William Gilbert, studied lodestones in what was the first sustained, systematic experimental investigation in the long history of *physica* (*De Magnete*, 1600). Fascinated by the magnetic compass that enabled the trade and piracy of London, Gilbert invented instruments to measure the dip and declination of a magnetic needle moved around a spherical lodestone, which performed like a little Earth, magnetically speaking. His method of arming magnets to increase their power instructed Galileo, and his premise that the Earth was a big magnet inspired Kepler. In vituperation of his opponents, 'intoxicated, crazy, puffed-up... lettered clowns', he outdid them both.

A final example, Simon Stevin, made his living as an engineer and state administrator after a late matriculation, in 1583 at the age of 35, at the University of Leyden. Stevin served the United Netherlands as military engineer and quartermaster. Although, unlike Galileo, Stevin made practical devices that worked, they took up many of the same theoretical problems and arrived at similar results, for example, the composition of velocities and the strength of gravity on inclined planes. Stevin declared for Copernicus publicly in 1608, before Galileo did, and proposed a method to find the longitude at sea using magnetic data no more practical than Galileo's scheme using eclipses of Jupiter's moons. In short, Stevin wrote on almost the entire canon of mixed mathematics, and would have been more influential in his time, and better known in ours, had he not written it in Dutch.

Chapter 4
Second creation

Revolution or integration?

'New philosophy calls all in doubt | The element of fire is quite put out | The earth's lost and no man's wit | Can well direct him where to look for it'. John Donne thus recorded the tremors in a world of learning shocked by the discoveries of the Columbus of the heavens and of the incompetence of all ancient systems of thought, apart from the sceptical, to accommodate modern knowledge. For a time the philosophical fashion was scepticism, which, strengthened by the recovered sceptics' handbook of Sextus Empiricus, made common cause with its emotional opposite, the voluntarism of Ockham and Urban.

The throwing of all philosophy into doubt inspired three important schemes for placing *physica* on firm foundations or limiting its claims to secure natural knowledge. By the middle of the 18th century, Newton's scheme had won out over those of Descartes and Leibniz, although only by reducing their rigorous logic to the easy-going 'reason' of the Enlightenment, and replacing their specifications of God's place in His creation by assimilating Him to Nature. Enlightened Newtonianism prepared the way for the invention of physics and its first standard model by overcoming the dichotomies between mixed mathematics and *physica*, and between celestial and terrestrial mechanics.

The appearance of an auspicious comet in 1577—which, according to parallax measurements, must have sailed through the planetary regions—underscored the need for a celestial mechanics that did without crystalline spheres. The supernovas of 1572 and 1604, likewise located beyond the Moon, disclosed that quintessence was liable to change. Tycho's sometime assistant and successor as mathematician to the Holy Roman Emperor, Johannes Kepler, not only freed the planets from their 2,000-year crystalline imprisonment, he also broke with all authority, ancient and modern, by holding that the planets employed their freedom by describing elliptical orbits with the Sun as a focus. He supposed that the Sun rotates so that its sweeping rays push the planets around in circles, and that a magnetic force deforms the circles into ellipses. Kepler's solar lighthouse anticipated Galileo's discovery of the Sun's axial rotation by several years.

A younger contemporary of Gilbert's at the court of Elizabeth I, Francis Bacon, who rose to become the chief lawyer in the land, indicted the school philosophy on capital charges. He did not attempt a new celestial mechanics, but proposed a new method of inquiry, a new purpose for *physica*, and a new institution for pursuing it. The method was to collect all the information available on a given topic, extend and correct it by experiment, winnow and systematize the results, and, by 'crucial experiments', decide between different theoretical alternatives. The new purpose was 'the improvement of man's estate'; 'knowledge is power', and *physica* without practical applications is not knowledge at all. Bacon's institution was a house of Solomon equipped with all the books, instruments, and materials needed to devise a replacement for Aristotle. This would not be the leisured amusement of free men, nor the narrow speculations of monks, nor even the bookish science of university professors, but a systematic aristocratic experimental enquiry directed to the exploitation of nature (see Figure 14).

Although the Jesuits, the schoolmasters of Catholic Europe, still insisted on Aristotelian *physica* as fundamental to their

14. Royal Academicians. Louis XIV's main minister, Jean-Baptiste Colbert, presents the Académie royale des sciences to its master. The unfinished Royal Observatory (completed in 1671) appears in the background.

philosophy and theology, most knowledgeable people knew by the time Bacon died in 1626 that it could not occupy its position much longer. Atomism offered a possible replacement. Galileo had endorsed it in his anti-Jesuit *Saggiatore* (1623) and Pierre Gassendi, a priest and professor in Paris, Christianized Epicurus as a paladin against Aristotelians and sceptics. Nonetheless, Gassendi ended near scepticism by allowing God the freedom to alter the order of nature. Eliminating scepticism and voluntarism to gain space for physics was the providential mission of René Descartes, whose famous bout of doubt ended with the assurance that God would not deceive him about any conclusion he apprehended with the same clarity and distinction as the inference 'I think, therefore I am' (*Discourse on Method*, 1637). This principle required God to have made necessary what Descartes conceived as such, and to stick to His original scheme; changing His mind merely to show His power would be childish. Descartes backed up these gratuitous assertions with spectacular acquisitions he claimed to have made by his method, which he exhibited in *Essays* appended to his *Discourse.*

One *Essay* presents his invention of analytic geometry—a proof, he said, that the ancients did not know everything about mathematics. A second, on optics, shows how to improve telescope lenses and how the lenses and muscles of the eye work. The third essay, on meteors, provides a corpuscular mechanism for most of the physical phenomena Aristotle treated in his *Meteorologica* and Pliny crammed into his *Natural History*. It also offers a singular result, obtained using the law of refraction, $\sin i = n \sin r$, now known as Snell's law after a professor at Leyden. Applying Snell's law to raindrops, Descartes calculated, from the known value of the index of refraction of water n, that the maximum height of the primary bow must be Roger Bacon's 42°.

Descartes' *Principia Philosophiae* of 1644 extends the corpuscular approach to the full spectrum of *physica*. It begins with laws of motion expressed mathematically, which, in principle, regulate the intercourse of the differently shaped bits making up the physical world. These bits are not atoms, as Descartes' horror of the void exceeded Aristotle's, and, despite his claims, they do not obey his laws of motion. Still, to make a universe out of mass, motion, shape, and the impossibility of a vacuum would challenge a Demiurge. Descartes met it, like another Timaeus, with a creation myth, which derives the current universe from an original undifferentiated space-matter activated by God's command, 'let there be motion'.

In this myth, the heavenly bodies, each of which once sat as a sun at the centre of its own vortex, can be swept into their neighbours' whirlpools. Earth thus captured Moon before both were torn into Sun's maelstrom. What the Earth–Moon system retained of its original vortex causes the fall of bodies, the tides, and the magnetic phenomena detailed by Gilbert (see Figure 15). Particles of various shapes, and bodies with special pores, account for all the activity in the visible universe including the machinery of the body. The only substance in the world besides the universal space-matter is mind, which, by a mysterious connection to the body at the pineal gland, supplies all the colour, sensation, and drama of life.

Descartes did not deliver his wide and whimsical world picture as true in all details. The creation myth does not agree with what Moses tells us of God's methods, and Descartes could not be certain which of the many mechanisms he could conceive God had chosen. Galileo had quantified a small portion of the science of motion in a manner Descartes criticized as spotty, unsystematic, and indulgent, whereas he, Descartes, had demonstrated that mathematical physics was possible and even useful. And, unlike Aristotelian philosophy, which only trained philosophers might think they understood, Descartes' spare system lent itself to development using intuitive concepts available to everyone who could think and read French.

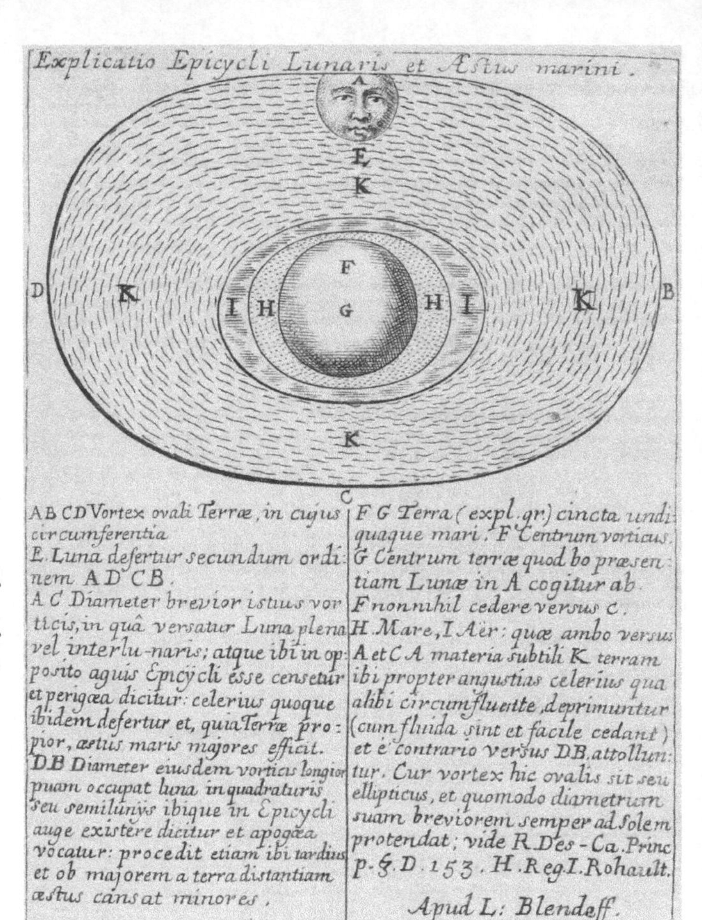

15. The Earth—Moon eddy in the solar vortex. The remnant of Earth's primeval vortex ABCD carries the Moon, and, flowing more quickly in the narrow passage between the two bodies, creates the tides.

The earliest cadre of Cartesians consisted of Dutch physicians attracted by the project of basing medicine on mechanical accounts of the body's functions and repelled by old-fashioned medical establishments. In France, lawyers at odds with a regime that rated ancient privilege above existing talent attended lectures by Descartes' disciples. One of their gatherings turned into the Académie royale des sciences in Paris (founded 1666). Also, many of the moving spirits behind the Royal Society of London (founded 1662), for example, Robert Boyle, had flirted with Cartesian philosophy. And so the creation of the first enduring academies of science coincided with the promotion of the first natural philosophy comprehensive enough to replace Aristotle's. Like the simultaneous appearance of Aristotle's *physica* and the universities some five centuries earlier, the incorporation of corpuscular philosophy in academies of science made a revolution.

The Scientific Revolution had an effective engine of war that Descartes had not thought of: the air pump. It debuted in 1654 in Regensburg, where its inventor, Otto von Guericke, Mayor of Magdeburg, brought it to amuse his fellow delegates to meetings called to clean up the mess left by the Thirty Years War (see Figure 16). The extraordinary power of nothing came increasingly under the control of experimenters, as Boyle, his sometime assistant Robert Hooke, and Christiaan Huygens (the leading member of the Paris Academy of Sciences) began improving Guericke's instrument. Spaces devoid of air were the first reliably reproducible artificial environments in which to experiment on nature.

The earliest such experiments, performed in the void space above the liquid in a barometer tube, date from the early 1640s when Galileans using a barrel, water, and a lead pipe investigated why ordinary pumps could not lift water higher than 30 feet. Galileo's prize student, Evangelista Torricelli, who succeeded him as Grand Ducal Mathematician, replaced the barrel, lead, and water with a small basin, glass, and mercury, and so created a compact instrument for measuring heights, tracking the weather, and testing the nature of apparently empty spaces. Such tests preoccupied the Accademia del Cimento, a short-lived group of Galileo's followers supported by the Medici between 1657 and 1667. They found that a void can transmit light and magnetism,

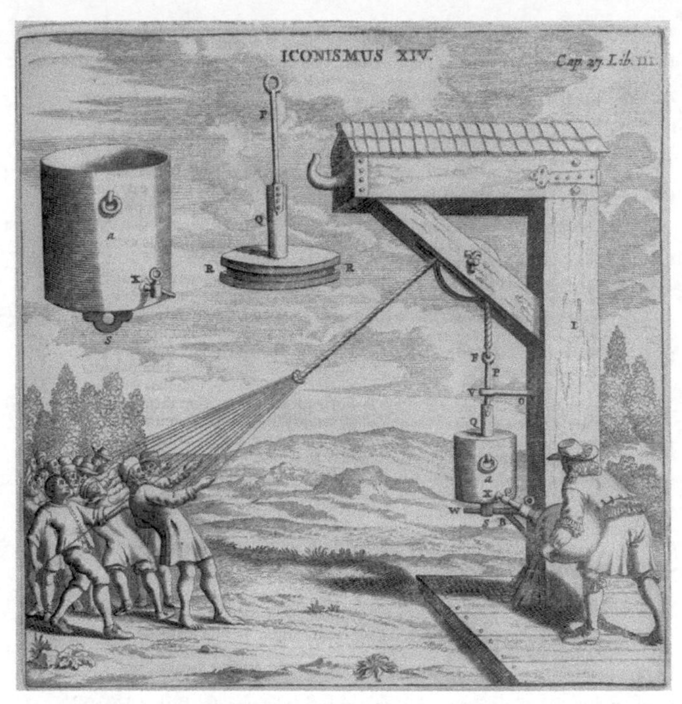

16. Diversion from politics. One of the demonstrations with which Otto von Guericke entertained delegates at Regensburg in 1654. The lid of the heavy cylinder cannot be removed by gravity or political maneuvering after the cylinder is evacuated.

but not sound and electricity. Evidently nothing has a capacity to discriminate.

Cartesian efforts to stave off the vacuum became less convincing after Blaise Pascal had explained that the barometer column stands because the atmosphere exerts pressure on the mercury in the basin and none in the empty space at the tube's top. Boyle and Hooke put numbers on the pressure by sealing a sample of air in a U-tube closed at one end and pouring mercury in at the other. In 1662, Boyle published their rare example of early quantification in *physica*: the air's spring is as its density (pressure $\propto$ volume^{-1}). Hooke found a similar relationship for a real spring: force is as extension, *ut tensio sic vis*, or, as originally formulated to secure priority and monopoly, 'ceiiinosssttuv'.

As revolutions should, the Cartesian immediately stirred up opposition. The curators of Dutch universities made common cause with Jesuit generals in prohibiting the teaching of any philosophy but Aristotle's. The Archbishop of Paris shut down public lectures on Descartes' physics, the Holy Office condemned his philosophy, the Index prohibited his works, and the Papal and Spanish authorities who ran Naples jailed doctors and lawyers who defended his ideas. When the smoke cleared around 1700, the enlightened Cartesian had become the prototypical academic in Paris. The career of the Oratorian Nicolas Malebranche was exemplary. When fresh from his scholastic training, he could not peruse Descartes without palpitations of the heart. Gradually he grew stronger and advanced from Aristotelian darkness to the light of mathematics, the harmony of sound physics, and a chair at the Academy. Malebranche's sound *physica* eased the problems of Cartesianism with the uneconomical doctrine of occasionalism, which considers pushes (the only source of physical change in Descartes' desert world) as occasions for God's continuous and ubiquitous action on His creation.

Mathematics or physics?

When Newton entered Cambridge as an undergraduate in 1661, the advanced thinkers there affected the philosophy of Descartes. Much of Newton's physics developed from an effort to grasp, demolish, and supersede it. He began the work in 1665–6 while rusticating at his mother's farm to escape a plague raging in Cambridge. There he encountered the legendary apple. Two decades later, while professor of mathematics at Cambridge, he put together the lesson from the apple with constellations of problems found in Descartes' *Principia*. The title Newton gave his masterwork reflects the inspiration he obtained from his predecessor's work and the main fault he found in it. Whereas Descartes' *Principia Philosophiae* dealt, like Aristotle's *physica*, with the foundation of all philosophy, Newton's *Principia Mathematica Philosophiae Naturalis* restricted his to the crucial elements of a mathematical physics.

Newton's first fruitful engagement with Descartes concerned the improvement of telescopes. It led to an experimental study of prismatic colours, which for its ingenuity, method, and care set a new standard in physical science. Newton modestly described its main result as 'the oddest if not the most considerable detection wch hath hitherto beene made in the operations of nature' when transmitting it to the Royal Society of London, then, in 1672, only a decade old. He had found that sunlight consisted of rays of different colours. Ingenious crucial experiments showed that each ray had a different and definite index of refraction.

Newton added to this kinematics of colour a conjectural dynamics in which rays of light consist of particles. Their release from incandescent bodies sends fast pulses through a space-filling springy ether. A ray encountering a transparent or translucent body crosses its surface or rebounds from it according to the phase of the ether vibration there. Neither Newton's chromodynamics nor 'the oddest detection' it explained pleased *physici* wedded to the hoary concept of colour as impure light. Among the loudest of Newton's critics was Hooke. While Newton vegetated on the farm, Hooke published his magnificent *Micrographia*, which, besides memorable detailed illustrations of a flea and other personal items seen through his microscope, discussed colour as modified white light.

The controversy with Hooke and others caused Newton, who preferred to dictate than to argue, to lay aside his work on light and colour until Hooke died in 1703. Newton then published a big book on *Opticks* (1704) and accepted the presidency of the Royal Society, which he ruled from 1703 until he died in 1727. The several editions of *Opticks* served as guides to the curators of experiments at the society via a growing number of 'queries' ranging freely through *physica*. The conjectural functions of ether (or several ethers) increased to cover gravity, magnetism, and electrical attraction, while other conjectures allowed the particles of bodies to interact directly through space. Whatever their plausibility, his conjectures, like Descartes' vortices, applied the same physical principles to celestial and terrestrial phenomena, and to the unseen world. This presumption—that principles adequate to explain the visible universe apply universally—which Newton stated explicitly as a rule of philosophizing, guided physical theory until the 20th century showed its untenability.

Newton's glory was to fulfil, in his *Principia* of 1687, Galileo's hope of geometrizing gravitation. The ingredients of his solution were Descartes' first law of motion (the principle of inertia), Galileo's rules of free fall and composition of velocities, and Kepler's rules of planetary orbiting. Newton showed that Kepler's rules followed from Galileo's, the principle of inertia, and the assumption that a planet falls towards the Sun along the line joining their centres. Since Kepler's rules allowed the substitution of an area for a time, Newton could reduce the problem of the magnitude of gravitational acceleration to a problem in geometry. He found that if the attractive centre is at a focus of the ellipse, the strength of the acceleration declines inversely as the square of the distance between the Sun and planet (see Figure 17).

According to the apple story, the idea of universal gravitation hit Newton when he perceived that the Moon should fall towards Earth's centre as an apple does. Assuming that the Earth acts on both the apple and the Moon as if its entire accelerative power were located at its centre (a theorem Newton later proved), he estimated the acceleration of the Moon and compared it with the amount required in his theory to keep it in circular orbit. The two estimates differed by one part in 3,600. It followed that if Newton could throw an apple hard enough he could turn it into a moon.

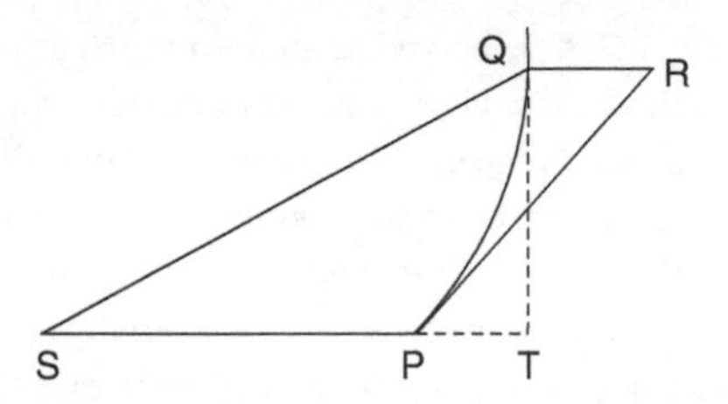

17. Planetary motion geometrized. Suppose the planet moves from P to a close point Q in the very short time *t* at a constant acceleration α toward the fixed center S. If instead it escaped from S's control at P, in time *t* it would have moved along the tangent to, say, R. By Galileo's rule, $\alpha = 2QR/t^2$; according to Kepler's second law, area SPQ $\approx SP \times QT/2$ is proportional to t; consequently $\alpha \propto QR/(SP \times QT)^2$. It requires only geometry and cleverness to show that if PQ is part of an ellipse with S at a focus, $\alpha \propto 1/SP^2$.

As they circle the Sun in their mutual embrace, Earth and Moon revolve around their common centre of gravity (see Figure 18). At Earth's centre, the Moon's attraction just balances the tendency to fly off on a tangent. On the side of the Earth nearest the Moon, therefore, the attraction exceeds centrifugal force; on the far side, falls short; and so the Moon causes two diurnal tides. The additional pull of the Sun accounts for springs and neaps. Newton thus quantified several of the enduring problems of *physica* that Descartes had bundled together: the nature of light and colour; the cause of the tides and the principles of planetary motion; and the size and shape of the Earth. The shape turned out to be only approximately spherical. The gravitational theory, applied to a spinning blob of molten rock (Earth in Newton's creation myth), brought forth a pumpkin, several miles longer in its equatorial than in its polar axis.

Unfortunately, the *Principia* was not physics. Or so said the Cartesian reviewer in the *Journal des Sçavans.* As Newton acknowledged, he had not assigned a physical cause to gravitation, but had adopted a mathematical fiction, an immediate action at a distance. From a Cartesian point of view, he had retrograded. Either he had reintroduced occult causes with which the schools

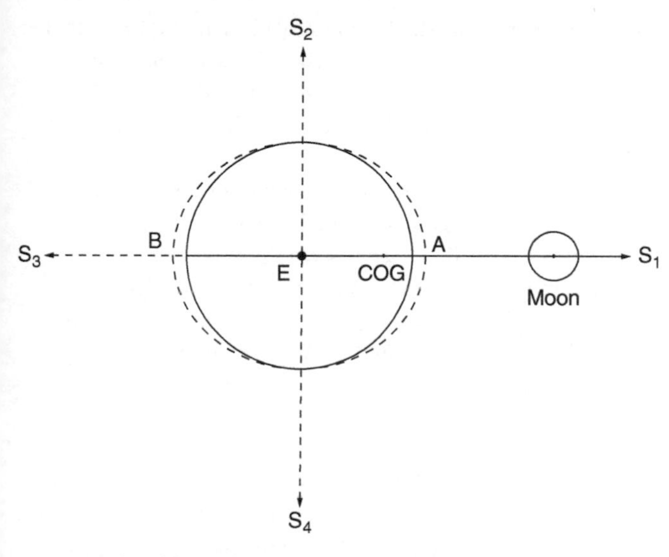

18. Newtonian tides. Newton's gravitational theory requires the Earth and Moon to saunter around their common center of gravity (CoG). Since centrifugal and gravitational forces balance at Earth's center E, the latter exceeds the former under the Moon and the former the latter opposite the Moon. The dashed oval represents the strength of the net vertical force, not a bulge in the sea; the driver of the lunar tides is the component of the net force parallel to the water surface. At S_1 (new moon) and S_3 (full moon) spring tides occur; at half moon (S_2 and S_4), neap tides.

avoided explanations or, taking at face value that he wrote merely descriptively of observables like accelerations, he had used the old dodge of separating mathematics from physics when the going got tough.

Newton replied to this criticism in an eloquent 'General Scholium' added to the second edition of the *Principia* (1713). He declared that the universe is not the work of an absentee mechanic, but proceeded 'from the counsel and dominion of an intelligent and powerful Being... eternal, infinite, absolutely perfect'. How the Being contrives gravity, His servant Newton did not know, and

‘I frame no hypotheses, whether physical or metaphysical, whether of occult qualities or mechanical’. Exact description sufficed. ‘[T]o us it is enough that gravity does really exist, and act according to the laws that [I] have explained, and abundantly serves to account for all the motions of the celestial bodies, and of our sea’. Thus Newton completed one full turn of the helix of scientific advance. His gravity resembled Aristotle’s, his omnipresent God the Stoics’ *pneuma*, and his planetary theory, with its preference for mathematics over physics, Ptolemy’s artificial circles. In his reversion to these elements of the *ancien régime*, in his system of laws and his efforts to impose them, and in his imperial presidency of the Royal Society, Newton was the Napoleon of the Scientific Revolution.

Newton aimed his carefully crafted epistemology against sceptics and *physici* who hesitated to grant that his system was free from hypotheses. For people like himself who worried that his system did not offer enough employment for God, he assigned the Creator the task of keeping the planets from falling into the Sun. This tinkering seemed ridiculous to Leibniz, who had fashioned his own response to sceptics and providentialists. His God, like Descartes’, did not behave arbitrarily; but whereas Descartes allowed God a choice of laws and initial conditions, Leibniz observed that His character had no room for caprice. It required that He make the best of all possible worlds.

The invention of physics

The demonstration of new *physica* to a fee-paying public via experiments had a vogue in Cartesian circles after 1670. The notable figure here, Jacques Rohault, described his offerings in a text that Newton’s spokesman Samuel Clarke thought proper to translate into Latin with prophylactic notes. Hooke’s successors as curator of experiments at the Royal Society during Newton’s presidency, Francis Hauksbee and the Huguenot J. T. Desaguliers, freelanced as public lecturers. Perhaps the most successful such performer was Jean-Antoine Nollet, a peasant's son who rose to be tutor to the royal children, lecturer at the royal artillery school, and a member of the Paris Academy of Sciences (see Figure 19). Beginning around 1720, experimental demonstrations of Newtonian physics entered classrooms, notably those of W. J. 'sGravesande and his successor Petrus van Musschenbroek in Leyden. Their textbooks defined the scope of experimental natural philosophy (or *physique expérimentale*) around 1750.

It was far narrower than *physica*. To attract a discriminating public, its demonstrations had to be clear, clean, and easily visible. These requirements, through which, unusually, the wider society helped define the *content* of physics, limited the subject to mechanical principles and model machines, pneumatics, optics, capillary and other hydrostatic phenomena, heat, magnetism, electricity, and astronomy as presented through the orrery (a mechanical solar system invented around 1704), armillary, and globes. After the accidental invention of the Leyden jar or condenser in 1745, electricity became the main attraction in this repertoire and Nollet its main expositor. His dominance did not last long. At mid-century, after exposure to sparks and glows by an itinerant lecturer in Philadelphia, Benjamin Franklin invented the system of positive and negative electricity. When interpreted as centres of Newtonian forces, Franklin's electrical charges drove Nollet's Cartesian theory of electrical matter from the field.

Electricity was also the bellwether in the quantification of physics. Charles Augustin Coulomb, an engineer trained at the artillery school in which Nollet taught and eventually a Parisian academician, devised an instrument that balanced the force between two electrified pith balls against the measurable torsion of a wire. Although few people if any have been able to reproduce his experimental results, they persuaded colleagues ready to believe that the force between hypothetical droplets of electric charge diminishes, as does gravity's, as the square of the distance. Coulomb did similar measurements on long bar magnets and got

19. Fashionable lecturing around 1750. The lecturer, the abbé Nollet, whose clientele included the royal family, is charging a human condenser.

similar results: droplets of the same magnetic fluids repel, and of different ones attract, one another via an inverse-square force.

Coulomb's experiments date from 1785. By then other *physiciens-géomètres*—F. U. T. Aepinus (electricity and magnetism), Henry Cavendish (electricity, thermometry, pneumatics), Pierre Simon de Laplace (heat, optics, capillarity), to name but one prominent representative from each of the leading academies (Berlin and St Petersburg, London, Paris)—were quantifying the new domains of physics in analogy to gravitational theory. One of these domains, which, after 1770, rivalled electricity in public demonstrations, was pneumatics. It expanded following the discovery by Joseph Black, Joseph Priestley, and Cavendish that 'air', which had been considered elemental since antiquity, consists of distinct permanent types. This capital discovery—which was analogous in form, as it was equal in importance, to Newton's detection of the composite nature of sunlight—created a productive overlap between physics and areas subsequently assigned to chemistry. The overlap widened around 1800 when the invention of the battery by Alessandro Volta (a Catholic professor in Pavia), and of a way to associate numbers with atoms by John Dalton (a Quaker school teacher in Manchester), demonstrated both the ecumenicity and accessibility of the then nascent physics.

The principles underlying the main branches of physics around 1800 bore analogies strong enough to encourage the belief that science might regain the sort of unity that Aristotelian concepts once gave *physica*. A 'standard model' emerged that comprised two electric and two magnetic fluids, Newton's light particles, and caloric, a self-repellent fluid of heat that accounted for thermal phenomena and the elasticity of gases. In addition, incorporating ideas developed by Jean-André Deluc (a Genevan merchant who became tutor in natural philosophy to the Queen of England) and others who studied nature's grand open meteorological laboratory, the model emphasized the interactions of caloric, light, and electricity among themselves and with the gas types. The special qualities of the active fluids, all conceived as weightless centres of distance force, displayed strong analogies, such as those between temperature and electrical tension, and between capacities for heat and electricity. The 'imponderable standard model' thus had a plausible coherence, although through form rather than matter.

The dominant imponderable model had a few competitors. One, explicitly opposed to the atomic picture, traced its roots to Immanuel Kant's *Metaphysische Anfangsgründe der Naturwissenschaft* (1786), which constituted matter and its properties from opposing forces of attraction and repulsion. By almost pure thought F. W. J. Schelling developed a *Naturphilosophie* that extended the scheme of polar forces to all physical phenomena. *Naturphilosophen* less doctrinaire than he, like H. C. Ørsted (Copenhagen) and Thomas Seebeck (Berlin) sometimes entered the laboratory where, guided by notions of the unity and polarity of matter, they discovered connections between electric currents and both magnetism (Ørsted) and heat (Seebeck).

Institutional frameworks

The early-modern college or university, which catered to students in their early teens, disseminated rather than advanced knowledge. Teachers of physics ambitious for literary fame would write a textbook, not a research paper, and, if they showed experiments, provided their own instruments. Around 1750, however, universities and colleges began to acquire the apparatus of deceased professors at a discount and soon had the burden of maintaining and augmenting their bargains. The cabinet of instruments, with its director-professor, mechanic-servant, and small budget, thus became as much a part of a university as a library. Some research inevitably resulted, notably at Bologna and Pavia—showplaces, respectively, of the Catholic Church and Austrian Lombardy—and at Protestant Göttingen, the only significant university founded during the 18th century in Europe. Physics as a career began to take shape: in the early 18th century three out of four professors of physics left their chairs, usually to teach and practice medicine; in the late 18th century, the proportion was reversed, and most professors of physics retired from or died in harness.

Some of these persevering professors worked in newly established higher schools for engineering, mining, forestry, and the military, such as the royal artillery school in Mézières where Nollet taught and Coulomb studied. Fifty years later, the then new École polytechnique, a product of the educational philosophy of the French Revolution, taught the latest quantified physics and prepared the leaders and professors of physical science in France. The tie between teaching in these technical schools and the systematization of physics may be symbolized by the four-volume *Traité de Physique Experimentelle et Mathématique*, the first modern textbook of physics, published in 1816 by a graduate of the École polytechnique, Jean-Baptiste Biot.

The advancement of knowledge was the mission of academies. Their self-image may be inferred from their mottoes. The Royal Society's ran *Nullius in verba*, a contraction of Horace's 'bound to swear in no master's words'. Thus they declared independence from their king, the tenets of the schools, and the dogmas of religion. The Paris Academy at its reorganization in 1699 took as its seal a sun, symbolic both of knowledge and of its king, Louis XIV, and for its motto, *Invenit et perfecit*. Its business was discovering and completing knowledge. The Royal Swedish Academy of Sciences, established in 1739, preferred to identify with an old man working the earth under the slogan, *För Efterkommande*, 'for posterity'.

Though dissimilar in structure and operation, the prototypical academies of Paris and London had in common the pursuit of natural knowledge exclusively. Most other academies also had sections for arts and *belles-lettres*. At least sixty such academies with substantial interest in natural knowledge were founded between 1700 and 1790. France led the way, with half the total, reflecting a need felt by lawyers, priests, physicians, and military officers to express themselves outside the institutions of church and state. These academies helped to guide research in physics by judging papers for publication and awarding prizes and medals.

In contrast with astronomy—which, with the introduction of achromatic lenses around 1760 and the construction of large reflectors in the 1780s, produced discoveries of a quality not seen since Galileo (notably William Herschel's planet Uranus)—physics experiments typically did not take place in facilities specially built or equipped for the purpose. Most experimenters adapted rooms in their homes, and household implements, to their purposes. Sometimes they enlisted workshops, as William Lewis did to study platinum. Count Rumford (the American Tory) famously discovered in the grinding of gun barrels a powerful argument in favour of the kinetic theory of heat. Another quasi-industrial setting for experiments was the instrument-maker's shop.

Among reserved spaces for physics experiments, Martinus van Marum's deserves special mention. Practicing physician, public lecturer, curator of instruments, and academician, he had charge of a rich private institute for the cultivation of arts and sciences (the Teyler Foundation in Haarlem). There, implementing Priestley's belief that the bigger the apparatus the better the discoveries, van Marum commissioned the largest electrostatic generator then yet made. It did not repay the investment.

The great Teylerian electrical machine was the king of a line of lesser devices that, like it and the air pump, could serve both research and demonstration. Measuring devices, in contrast, were unlikely to entertain a crowd, and at first did not enlighten their users. They became more apt and exact after 1770, with Volta's electrometers and standardized accurate thermometers, barometers, and hygrometers developed from work by Deluc, Cavendish, and others. Routine meteorological and geophysical measurements inspired the invention of reliable recording instruments. The subtlest recorder employed in physical experiments in the 18th century was a frog. Using decapitated specimens Luigi Galvani (Bologna) detected the animal electricity that inspired Volta to make electrometers better calibrated than amphibians, to generate animal electricity without using animal parts, and to magnify the effect into an epoch-making prime mover: the Voltaic pile or battery.

Animal electricity or galvanism made spectacular demonstrations. People enjoyed watching Guillotined heads wired to a Voltaic cell wink. Other responses were coaxed from the hanged murderer to whose spinal cord and muscles Dr Andrew Ure of Glasgow attached galvanic leads. The corpse showed rage, agony, and despair, and died again with a horrible smile.

Physics and enlightenment

Several factors operating in the wider society reinforced the rationalization of natural knowledge around 1750. One was the increasing application of mathematics to the purposes of the enlightened governments of the later 18th century. Another was manufacture, as represented particularly by the trade in instruments. The *Encyclopédie* of Denis Diderot and Jean Le Rond d'Alembert, published for the first time between 1751 and 1772, made much of the rationalization of the manufacturing arts. Towards the end of the century, rhetoric connecting physics and utility could emphasize the spectacular cases of the lightning rod and navigable balloons.

The French Enlightenment drew on rationalized natural knowledge for its central project of destroying organized religion. Voltaire's *Elémens de la Philosophie de Neuton* (1738), the first comprehensive popular account of Newton's work on cosmology and optics in French, begins with Newton's notions of God and proceeds to correct thinking about free will, natural religion, and mankind's common humanity. Francesco Algarotti added enough of the philosophy of John Locke to his *Newtonianismo per le Dame* (1737) to excite the Congregation of the Index. Pope Clement XII and the Holy Office had then just condemned two of Locke's major works. The Catholic Church thus helped to recommend Locke's sensationalist psychology as a weapon against itself in the *philosophes*' campaign to crush it.

D'Alembert coupled Newton and Locke, the one the reformer of physics, the other that of metaphysics, in his lengthy 'Discours Préliminaire' to the *Encyclopédie* (1751). Newton had demonstrated how far the human mind could soar when freighted only with mathematics and a few capital phenomena, and Locke had shown that all minds begin with even less, that is, with nothing at all. David Hume then worked hard to show that whatever the mind acquires, it will not get to the bottom of things.

Kant dug a deeper foundation for the proposition that the human mind cannot penetrate to things as they are. What the mind can acquire dependably is knowledge only of the appearances it arranges in space and time and orders, where appropriate, by cause and effect. Among the conclusions that Kant's mind derived from these principles was that physics must be quantitative: 'in every special doctrine of nature only so much science can be found as there is mathematics in it'. Hence, he wrote three years before the publication of Lavoisier's *Traité de Chimie*, 'Chemistry can become nothing more than a systematic art or experimental doctrine, but never science proper'.

The Enlightenment's programme of rationalizing social institutions and artificial practices, and of sweeping away customs and concepts that had no deeper foundation than habit and tradition, coincided with the interests of bureaucratizing governments eager to curtail wasteful inefficiencies. The tougher, leaner, more mathematical experimental physics of the later 18th century both served and profited from this coincidence.

Particularly after the Seven Years War, the larger states recognized the need for a better accounting, and more efficient use, of their resources. The entrance of the expert, the first step towards realizing the rationalization demanded by Enlightenment, the beginning of the effective mathematizing of natural philosophy, and the first glimmers of manufacture of items with interchangeable parts, were coeval.

After making his reputation in Bavaria designing nourishing soups and smokeless fireplaces, Count Rumford inspired the creation of an institution to transmit the latest science of soups, chimneys, and manure, the Royal Institution, which opened in London in 1799. It was to make capital contributions to natural philosophy if not to scientific farming. Its second professor of natural philosophy was Thomas Young, reinventor of the wave theory of light. Its professor of chemistry, Humphry Davy, obtained funds for a very large Voltaic battery on the grounds that electricity enhanced the fertility of soil. With its aid, he separated sodium, potassium, and calcium from their oxides.

'Science' in the 18th century often meant rational cut-and-try. It had its success in agriculture and among the inventive members of the informal Lunar Society of Birmingham, who met when the full Moon illuminated their travels. Most of them also belonged to the Royal Society. The best known of their productions was the improvement of the steam engine by its member James Watt, who had learned something about heat and experimentation when serving as curator of instruments at the University of Glasgow. The society thus supplied an early example of the academic–industrial complex as well as the machine that would propel the Industrial Revolution.

Because of the emphasis on mathematics in their curriculum, Jesuits sometimes occupied important posts for which it was requisite. For example, Ruđer Bošcović (Roger Boscovich), a mathematics professor in Rome and later in Pavia, helped in repairing the dome of St Peter's, made a trigonometric survey of the Papal States, founded the Jesuit observatory in Milan, and acted as director of optics for the French navy. He is best known, however, as the inventor of an extreme Newtonian natural philosophy based on a single universal distance force (*Theory of Natural Philosophy*, 1758, 1763).

The welter of German states and principalities, which abounded in universities and bureaucracies, and, after the Seven Years War, in problems of reconstruction, provided a theatre for a special combination of economic rationalization and natural knowledge. *Cameralwissenschaft* collected and integrated information about agriculture, commerce, industry, population, territory, history, laws, and customs to serve as a basis for economic policy. Perhaps its most successful and demanding subject was forestry, devoted to increasing sustainable yields by planting and harvesting trees in accordance with rationalized experience and the integral calculus.

Judging the health of the state by reading the 'thermometer of public prosperity', that is, the census, brought natural philosophers close to the people, especially in France. Owing to the size and diversity of the population and its fear that a census would reveal its prosperity, numbering the king's subjects was not easy. The confident new natural philosophers and mixed mathematicians did not think it beyond reach. 'Experiment, research, calculation are the probe of the sciences! What problems could not be so treated in administration! What sublime questions could not be submitted to the law of calculation!'

Chapter 5
Classical physics and its cure

During the 19th century, physics became weightier in several senses. The number of imponderables fell almost to zero. Physics became a recognized profession and its practitioners 'physicists'. It and they acquired special training facilities in university institutes and technical schools that sprang up in Europe and America after 1870 primarily to train students as schoolteachers and electrical engineers (see Figure 20). At the same time, the perpetually shifting border between physics and mathematics moved to make room for theorists; and again, as happened with the discovery of the gas types, fertile concepts, such as the ion and the table of elements, came over from chemistry.

Nineteenth-century physics ended with a new standard model or project more Cartesian than Newtonian. The new project penetrated more deeply than its predecessor, which had obtained superficial unity with similarly acting, but distinct imponderables; whereas it premised an ultimate reduction to the same matter in diverse modes of motion. Once again Descartes did battle with Aristotle, accomplished a mighty synthesis, and prepared the way for something new. The physicists of the early 20th century admired, and set aside, their 19th-century accomplishments as 'classical', and built a quantum science of atoms and molecules that, they claimed, contained all of physics, and chemistry too, 'in principle'.

20. A physics institute around 1900. The size, fit with local architecture, and arrangement of lecture halls, laboratories, and utilities at the University of Manchester were typical of institutes built around 1900. Rutherford was professor and director there from 1907 to 1919.

From around 1850, physics increasingly made good on Bacon's promise that experimental science would improve the human condition. New industries competed to supply electric communication, light, and power, and to invent electrical appliances for the home and the workplace. New governmental agencies, national and international, regulated services, and standardized products. Old universities introduced new curricula to compete with technical higher schools and replaced academies as the primary loci of research. And societies of physicists sprang up to see to the professional interests of their members.

Standard models

Newton's rays of light had had a rival since 1690, when Huygens explained double refraction by assuming light to be a pressure wave in a world-filling ether. The first wave theory competitive with Newton's ray theory did not appear for a century. Then around 1800 Thomas Young, having earned a medical degree in Germany with a thesis on sound and hearing, analogized the medium supposed to transmit light to a thin air capable of supporting acoustic vibrations. Thus he modelled the interference patterns of light rays passing through two parallel narrow slits as constructive and destructive combinations of sound waves. His model met resistance in a phenomenon foreign to sound: polarization.

The need to capture polarization became urgent when French physicists, expecting to refute Young's theory, unexpectedly confirmed an apparently absurd consequence of it. The absurdity, discovered by Augustin Fresnel, a graduate of the École polytechnique, was that the patterns produced by light falling on a screen through a minute hole could have dark blots at their centres. Young and Fresnel then independently suggested that the light-bearing medium could account for polarization if it oscillated at right angles to the direction of the wave. To transmit these 'transverse' vibrations, the 'luminiferous ether' had to behave like an elastic solid rather than a thin air. Physicists fatigued themselves during the rest of the century imagining ether rigid enough to propagate light yet soft enough to pass the planets. Nonetheless, most knowledgeable people had dropped the light particle in favour of ether vibrations by 1830.

At the same time, and following a similar pattern, the magnetic fluids also disappeared. Again an experimental demonstration made on the periphery (Paris then being the centre of physics) prompted a French mathematician to a destructive generalization. The actors were H. C. Ørsted (Copenhagen), who discovered that a current-carrying wire could act at a distance on a magnetic needle (1819), and A.-M. Ampère (Paris), who showed that a circular current behaves as a magnet and deduced that magnetism arises from electricity in motion (1822, 1827). This model implied the existence of distance forces as awkward as an elastic-solid vacuum: the magnetic force of a current lies not in its direction, but in circles around it, and the force between moving droplets of electrical fluid depends on their relative velocities as well as on their separation. Furthermore, to include the generation of an electric current by a changing magnetic force (1831), the most famous of the many famous discoveries of Michael Faraday (Royal Institution, London), the force-law must also involve the relative acceleration of the droplets.

In an ironic reversal of polarity, the proponents of these peculiar distance forces were continental physicists, notably Wilhelm Weber (Göttingen) and Hermann von Helmholtz (Königsberg, later Bonn and Berlin), whose predecessors had regarded Newton's gravity as unintelligible, whereas Newton's countrymen, Faraday and his followers, tried to do without distance forces altogether. Faraday located electricity in a special medium he supposed to exist in the space between electrified bodies. The stresses and strains in this medium or 'field' constituted, and conveyed, what appeared to ordinary physicists as forces acting at a distance between 'charged' or 'magnetized' bodies. William

Thomson, later Lord Kelvin (Glasgow), translated Faraday's intuitions into a dynamical picture portraying magnetic forces as vortices, and electric forces as linear flows, in the field. In the 1860s James Clerk Maxwell (Aberdeen and London, later Cambridge) worked out a dynamical model of Faraday's field capable of representing most known electrical and magnetic phenomena. On this expendable scaffold, he raised the edifice of an enduring electrodynamics, a set of relations ('Maxwell's equations', 1866) linking electric and magnetic forces and their sources. For those who followed the Faraday–Maxwell line, electrical fluids had no better claim to existence than magnetic fluids or corpuscles of light.

Nothing, therefore, could have been more gratifying to the grand-unifying physicist than to learn that Faraday's field and the luminiferous ether were one and the same. In 1864, Maxwell calculated from known electric and magnetic parameters that the speed with which the electromagnetic disturbances predicted by his model travel through space came close to the latest measurement of the speed of light. In 1887, Helmholtz's former student Heinrich Hertz (Karlsruhe) generated and detected the electromagnetic waves Maxwell had foreseen. The resultant grand synthesis complicated the task of ether-field theorists, who now required an elastic-solid medium filling all space, competent to produce all the phenomena of electricity, magnetism, and light, and transparent to gravitating bodies.

Like light particles, caloric had had a challenger in the 18th century that supposed heat to be a mode of motion. Count Rumford's production of heat from grinding cannon muzzles, though pertinent to the challenge, had not won it. The decisive experiments against the caloric theory, which date from the 1840s, also took place in an industrial setting, a brewery whose owner, James Prescott Joule, controlled his product by exact thermometry. Joule had studied with John Dalton and knew how to improve his leisure. Applying his instruments to engines, he showed that they consumed heat in raising a weight, and that a certain quantity of mechanical effort, or a certain expenditure of zinc in a battery, always generated the same amount of heat. Unlike matter, heat could be created and destroyed! Joule's conclusion had to overcome not only scepticism about his thermometers but also a brilliant theory of steam engines.

This theory, which applied to all engines, regardless of fuel, mechanism, and working substance, assumed the indestructibility of caloric. Its author, Sadi Carnot, a *polytechnicien* like Fresnel, calculated the maximum efficiency possible for any heat engine as a function of the temperatures of the boiler and condenser. To escape a possible perpetual motion, the most efficient machine must be reversible; and to be reversible, its operations must never allow contact between parts maintained or working at different temperatures. From the ideal cycle of operations he invented to satisfy this condition, he deduced that the engine's efficiency, the ratio of the weight lifted to the amount of caloric fluid Q employed, should be proportional to the difference of temperature ΔT between the boiler and the condenser.

The clarity, ingeniousness, and plausibility of Carnot's analysis of 1824 impressed William Thomson. But he also inclined to Joule's theory, which, in contrast to Carnot's, destroyed caloric in doing mechanical work. Mathematics indicated a path to reconciliation. Thomson saw that Joule's expression for efficiency would be the same as Carnot's if the conserved quantity was not Q, but Q/T. Two principles, not one, were involved: (1) mechanical work and electricity could be converted into heat and vice versa without loss (Joule); and (2) a heat engine working reversibly conserves the quantity Q/T (Carnot).

Rudolf Clausius (Bonn) also reconciled Carnot's conservation theory with Joule's transformation theory, which Clausius had in the form developed independently of Joule by the physician Robert Julius Mayer. Clausius called the coin of conversion among the forces of nature 'energy'. The mysterious quantity Q/T—nature's fee for the conversion—he called 'entropy'. In the ideal reversible case, the fee is zero. In practice, small amounts of heat are lost so as to render all 'real' processes irreversible. In Clausius' formulation, the energy of the universe stays constant while its entropy strives for a maximum ('the laws of thermodynamics', 1865). Thus considerations arising in industrial settings drove the most plausible and ancient of the imponderable fluids, the matter of fire and heat, out of physics.

The concept of heat as motion proved as far-reaching as the synthesis of light and electrodynamics. It made possible a quantitative link between mechanical concepts and temperature via parameters that measured molecules and allowed mathematicians passage between the macro and microworlds. The wormhole to this wonderland lay in the equation for a 'perfect gas', a legendary medium described by Boyle's law as generalized by Laplace's protégé J. L. Gay-Lussac (Paris): pressure = (R × temperature)/volume. The constant R was the wormhole. A simple model ('the kinetic theory of gases'), in which the pressure exerted by a standard number of molecules N arises from their bombardment of the walls of a cubical container, gave an astonishingly simple 'law': the kinetic energy of a perfect gas molecule is $3kT/2$, where k, the bond between mechanics and thermodynamics, and the measurable and the molecular, is R/N.

This simple theory ignored collisions among gas molecules, distributions in velocity, mean free paths, and, most importantly, a statistical treatment of equilibrium. Clausius, Maxwell, and Ludwig Boltzmann (Graz and Vienna) added these realistic touches, and, after heroic calculations, recovered the old result: each degree of freedom of motion of a molecule has an energy $kT/2$ at equilibrium ('the equipartition of energy'). Molecules able to move only in translation in three dimensions have an average energy of $3kT/2$. The specific heat of a gas sample each of whose

N molecules has f degrees of freedom would be $(f/2)R$. The formula worked well—for monatomic molecules.

Heat equilibrium occurs when entropy is a maximum. This simple statement hides a deep difficulty. If a gas is a perfect mechanical system, its motion should be reversible. Hence a statistical-mechanical representation of entropy appeared impossible. Boltzmann countered that, owing to the colossal number of molecules in play, departures from equilibrium will almost certainly be reversed instantaneously (1872, 1877, the 'H-theorem'). Entropy S consequently could be understood as a function of the probability W of finding the system in a particular state. The wormhole k, rechristened 'Boltzmann's constant', links not only the macro and the microworlds, but also the living and the dead. It appears on Boltzmann's tombstone in the form $S = k\log W$.

The successes of the gas theory and Maxwell's electrodynamics, the equivalence of all energy to mechanical energy underwritten by the first law of thermodynamics, and the relative ease with which physicists reasoned with the intuitive concepts of matter and motion combined to return 19th-century physicists to the dream of Descartes. The president of the French Physical Society, Alfred Cornu (Paris), opened the first international congress of physicists, held in Paris in 1900, with the reassurance that Descartes 'hovered' over them. The great mathematician Henri Poincaré (Paris) gave the keynote address. He advised his auditors against developing a fondness for a revised-Cartesian or any other world system. Rather, they should collect the facts of experiment and arrange them for consultation in the most convenient manner. A good physicist was more librarian than philosopher.

Poincaré's unflattering estimate of the epistemological status of physics corresponded to the considered opinion of many physicists. Although they might act and speak as if they sought the truths of nature, when they philosophized they acknowledged that the goal was beyond them. A prime impulse to this neo-scepticism was the demotion, by Gustav Kirchhoff (Heidelberg), of the most secure branch of physics, analytical mechanics, from a true account of matter and motion to a mere description of it. This 'descriptionism' turns up in the influential epistemologies of physicists who took a broad view of their discipline, notably Ernst Mach (Vienna), and also Joseph Larmor (Cambridge), who sponsored an English translation of Poincaré's writings.

Physicists as librarians

The metaphor of the library suited a large part of 19th-century physics, which boasted many new laws or effects easily entered into Poincaré's imaginary catalogue. Representative entries concerning metals might read, alphabetically: 'Electric current, generated by heating a junction of dissimilar metals', Thomas Seebeck, 1822; 'Heat capacity, atomic, inversely proportional to atomic weight', Pierre Louis Dulong and Alexis Thérèse Petit, 1819; 'Heat conductivity, proportional to *T* times electrical conductivity', Gustav Wiedemann and Rudolf Franz, 1853; and 'Paramagnetism, proportional to $1/T$', Pierre Curie, 1895. The inverse of these effects also had entries, for example, 'Heat developed by an electric current passing a bimetallic junction', J. C. A. Peltier, 1834.

The library added new shelves with the discoveries that mechanical forces could create electricity (piezoelectricity) and magnetism (magnetostriction). Pursuing such reciprocal effects, James Alfred Ewing, who established the frontier between physics and engineering at Cambridge, discovered hysteresis, the lagging of magnetic effects behind their immediate causes (1882–5). The most exact entries in the library of physics around 1900 were the characteristic wavelengths of the series of spectral lines emitted by the elements when heated sufficiently. The first such survey, made by Kirchhoff and Robert Bunsen (Heidelberg) in 1860, employed salts vaporized in Bunsen's burner, which immediately revealed cesium and rubidium, so named after the strongest colours in their spectra. Spectral analysis opened a way to explore the constitution of stars as well as of substances here below. Librarians of spectra invented several cataloguing rules, notably the 'Balmer formula' for hydrogen (after Johann Balmer, 1885), and its generalization, by Johannes Rydberg (Lund, 1888), to many series and elements. Their formulations contained the clue to what the Greeks would have rated an oxymoron: the internal structure of atoms.

Meteorologica provided many important new entries for the physics catalogue of 1900. Dalton had endowed atoms with different sizes to explain why the atmosphere's constituents do not separate out according to weight. The different sizes, which implied different weights, cause each gas type to act as if the others were not present. After much wrangling over atomic weights, chemists came to agree enough in the 1860s to prompt the creation of a Ouija board that ordered known elements into families by weight and predicted the existence of unknown elements soon discovered. In the 1890s, meteorology again intervened significantly in atomistics when Lord Rayleigh (formerly Cambridge) discovered a minor constituent of the atmosphere he named argon. A family of 'noble gases' of the inactive argon type beginning with helium soon disclosed itself to the chemist William Ramsay. The new family fitted into the Ouija board perfectly, with one exception: argon and potassium had to swap places rightfully theirs by weight to preserve chemical periodicity. The noble gases thus confirmed the power, and deepened the mystery, of the periodic table of elements.

Further to geophysics, Newton's tidal theory, refined by 'harmonic analysis' (William Whewell, Cambridge), grew powerful enough to become seriously misleading. The prevailing theory of the Earth's formation around 1850, the 'nebular hypothesis' started by Kant and Laplace, derived our solar system from a spiralling gas cloud that heated as it coagulated and left Earth with a molten core.

Kelvin observed that a mass of liquid under Earth's surface should show tidal effects. None being known, he concluded that Earth is a rigid body.

From thermodynamics and the properties of surface rocks, Kelvin calculated that Earth has been cooling for more than a hundred million years—an indigestibly long time for people who credited biblical chronology, but not long enough for Darwin's evolution to work. More acceptably to biblical literalists, Kelvin did not allow the human race very long to improve itself. The second law, he asserted, insured that the Earth, once too hot for humans, soon would become too cold for them. His widely accepted views about the age and rigidity of Earth, apparently anchored on rock-hard physical principles, did not survive long into the 20th century. Analysis of seismic waves from deep earthquakes showed that only slow longitudinal waves crossed the core. This information allowed Beno Gutenberg (Göttingen, 1912) to estimate that the core's radius is almost half Earth's, and everyone to infer that, since liquids do not support transverse waves, Earth is not solid throughout. By similar techniques, Andrija Mohorovičić (Zagreb) located a discontinuity some 30 miles beneath the continents that divides the crust from the 'mantle' (1909–10). Thus macroscopic physics softened Kelvin's rigid Earth. As for Earth's age, microphysics identified an additional heat source in radioactivity that exploded it beyond the needs of evolutionists.

Kelvin's clouds

Kelvin was an exemplary physicist not only because of the importance of his discoveries, his breadth of interest, and his dedication to the programme of mechanical reduction, but also because he could diagnose serious difficulties in his science. In 1900, he bundled his reservations into two problem areas, which, in happy reference to meteorology, he labelled 'clouds'. One he seeded on the fact that no one had managed to design ether that could do everything Maxwell required of it. The other was the equipartition of energy.

Like Kelvin, Boltzmann acknowledged that the progress of mechanical reduction had its difficulties, but saw no other way forward. He ridiculed the physical chemist Wilhelm Ostwald (Leipzig), who championed a science based only on energy and its transformations ('energetics'), for admonishing colleagues not to 'make unto thee any graven image, or any likeness of anything'. This commandment (Exodus 20:4–6) dissuaded physicists no more than it had worshippers of the golden calf. However some, like Max Planck (Berlin), though critical of energetics, stuck as close as they could to the two laws of thermodynamics. Still others thought to reverse the argument that gave priority to mechanics and substitute electricity or heat in its stead. Boltzmann associated these schisms with the *fin de siècle* attack on received canons of art, music, and literature. Everywhere, he said in 1899, classicism had its enemies. But for clarity, longevity, and productivity he would stay with what he, perhaps the first among mortals, called 'classical physics'—the science of clear mechanical models in space and time—and trust that some imperial Newton would dispose of equipartition.

Kelvin's clouds grew more thunderous after 1900. Through Maxwell's grand synthesis, the problem of how light propagates in a transparent body moving through stationary ether had taken on new urgency. Theory had explained some optical phenomena by supposing that moving matter drags a portion of the ether with it; but when the same theory required that it be pulled by spinning charged disks, it stubbornly remained quiescent. And yet moving ponderable bodies seemed to drag the ether surrounding them, since experiments to measure their velocity through it invariably failed. The most famous and delicate of these experiments, by A. A. Michelson (Cleveland, later Chicago) and Edward Morley (Cleveland), date from the 1880s.

As when Galileo sliced through the accumulated conundrums of motion by replacing physics with mathematics, so now H. A. Lorentz (Leyden) transformed Maxwell's equations for moving bodies so as to kill terms that predicted detectable effects arising from motion. His manoeuver in effect did away with the ether as a mechanical entity. In 1905, Albert Einstein, then a patent examiner in Bern, recognized that the Lorentz transformations were just what he needed to describe the physics in bodies moving with constant rectilinear velocities relative to one another.

Special relativity (Einstein's theory of 1905) is as democratic as equipartition since it effectively places all observers at rest within their own ether. Light consequently always propagates towards or away from observers in free space at the same speed irrespective of their states of motion. To this principle Einstein added the equally intuitive proposition that the laws of physics should be the same in all inertial frames; no physics experiment can tell you whether the passing train or your train or both are in steady motion. From these easily comprehensible beginnings Einstein deduced very bizarre consequences: time dilation and space contraction (clocks run slower, and sticks grow shorter, in 'moving systems' as seen from 'stationary' ones), and the equivalence of matter and energy. However disorienting the conclusions, the principles could be construed as a rational expansion of the principles of classical physics.

Discharging Kelvin's second cloud required a passage through irrationality. The problem that provoked the madness concerned the equilibrium distribution of radiant energy contained in an oven at constant temperature ('blackbody radiation'). This seemingly obscure problem had some industrial interest, as it related to standards of illumination, and the librarians of physics, foreseeing its solution, already had a place for it on the shelf between electrodynamics and thermodynamics. When ether was treated as a mechanical system, however, equipartition awarded every mode of ethereal vibration the same amount of energy, $kT/2$. As emphasized in 1900 by Lord Rayleigh, this democratic division could be catastrophic. If, like other vibrating systems, the ether had many more modes of vibration at high frequencies than at low, all the electromagnetic energy available to it should run into the ultraviolet and beyond.

To escape this predicament, Rayleigh and, more assertively, James Jeans (Cambridge) suggested that the process might take as long as the age of the universe, thus saving equipartition by postponing equilibrium to a time when no one was likely to observe it. That was not reasonable. Knowing nothing of Rayleigh's deliberations, Planck published in the same year, 1900, a theoretical formula for the blackbody spectrum that appeared to agree with experiment. Soon, however, measurements in the infrared negated the formula. Planck thereupon found one that worked. He concocted a theory for it that subverted the classical physics of Boltzmann from which he thought he had derived it. That was irrational. Planck's 'quantum theory' soon joined with other evidence that the microworld could not be described using the ordinary concepts of physics.

The microworld

William Whewell, who gave the world 'physicist' and 'scientist', supplied Faraday with 'ion' to refer to the electrified unknowns in motion in a working electrolytic cell. In his doctoral dissertation at the University of Uppsala in 1884, Svante Arrhenius identified ions with charged molecular fragments and claimed their presence in all solutions. His concept of dissociation proved its power also in the study of electrical discharges through dilute gases. During the 1870s and 1880s, physicists discovered 'cathode rays', which proceed invisibly in straight lines from the cathode of the discharge tube to cause its walls to fluoresce. (Whewell made 'cathode' and 'anode' for Faraday as well as 'ion'.) The obvious hypothesis, which supposed the rays to be negative gas ions repelled from the cathode, failed when J. J. Thomson (Cambridge) and others showed around 1897 that cathode-ray particles must have a ratio of charge to mass (e/m) about 1,000 times as large as that of the lightest electrolytic ion, hydrogen. Thomson inferred that the cathode-ray 'ion' (his word was 'corpuscle') represented matter in a state of complete dissociation.

The recognition of the corpuscle/electron, which re-established electric charge in British physics, capped a series of extraordinary discoveries associated with the gas-discharge tube. Late in 1895 Wilhelm Conrad Röntgen (Würzburg, later Munich) deduced from the glow of a phosphorescent screen some distance from a discharge tube generating cathode rays the existence of penetrating radiation of novel character (see Figure 21). He could not reflect or refract its rays or bend them with a magnet; but they made good photographs of the inside of a living human hand. He summed up his knowledge of their nature by calling them 'X-rays'. Henri Becquerel (Paris) found that uranium gave off rays different from Röntgen's. Marie Curie and her husband Pierre Curie (Paris) recognized in 1897–8 that thorium and the elements polonium

21. Desktop physics. W. C. Röntgen pictured in his laboratory in Würzburg in 1895, the year he discovered X-rays. Research spaces in the institutes were few and small around 1900, although many physics experiments required dependable electricity and good vacuum pumps.

and radium, which they discovered in uranium ores, also had the ability to radiate ('radioactivity').

Thomson and his student Ernest Rutherford established that all the new rays ionized the air, which enabled Rutherford to distinguish a soft ('alpha') and a more penetrating ('beta') component among the Becquerel rays. The e/m of the beta ray came out close to that of the cathode-ray ion. Thomson's inference that electrons might be the building blocks of matter had further confirmation in experiments performed in 1896 by Pieter Zeeman (Leyden, then Amsterdam) as elucidated by Lorentz. The 'Zeeman effect' (the splitting of spectral lines in a magnetic field) revealed that the electric oscillator supposed responsible for spectral emission had an e/m close to that of Thomson's corpuscle. Thus, around 1900, physics got its first elementary particle and half a dozen rays for which its library had no shelf mark.

Continuing his study of alpha particles, Rutherford, now a professor in Montreal (1902), and his chemist colleague Frederick Soddy uncovered the tendency to self-destruction of radioactive substances. Most of the many fleeting decay products subsequently detected did not fit into the periodic table. This information prompted the realization, by 1913, that the discrepancy at argon/potassium was systematic: the Ouija board worked correctly when ordered not by weight but by an integer beginning with unity at hydrogen and assigned on the basis of chemical properties. More than one radioelement could occupy the same cell in the table ('isotopy'): atoms differing in weight could have identical chemical properties. As a discovery, isotopy ranked in logic if not in depth with thermodynamics: in each case scientists realized that a single concept (atomic weight, energy) required another (atomic number, entropy) to provide an adequate description of the phenomena.

Shortly before the recognition of isotopy, Rutherford invented his version of the nuclear atom. It deviated from Thomson's market-leading model, in which electrons circulate within a positively charged space, in being unstable both mechanically and electrodynamically. That had ruled it out in its previous versions. But by 1911 the few bold spirits who had given up on ordinary physics in the microworld could think that Rutherford's atom was so bad that it might be a good place to look for quantum activity. Its advantages included, besides having classical physics against it, offering a simple representation of Z, the atomic number, as the charge on the nucleus, and explaining some details about the passage of alpha particles through matter.

Physicists entered the microworld definitively during the last few years before World War I, when their ability to assign exact values to the charges, masses, and numbers of atoms and electrons established their belief in them. The charge on the electron was the most valuable value: it produced, via measurable quantities, m, N, and k. In 1910, Robert Millikan (Chicago) gave the value $e = 4.891 \times 10^{-10}$ esu (electrostatic units) in his 'oil-drop experiment', a refinement of a technique using water invented in 1898 by J. J. Thomson. Planck's theory of blackbody radiation gave e as 4.69×10^{-10} esu, close to Rutherford's value from counting alpha particles, 4.65×10^{-10} esu.

Jean Perrin (Paris) confirmed these values from another direction by obtaining N from measurements of the dance of uniform tiny gumballs suspended in water between the tugs of gravity and osmotic pressure. Their dance, Brownian motion, arises from imbalances in the impacts of water molecules on them. In 1905, Einstein had calculated the mean displacement of the gumballs over time from the places at which they are first seen. The formula included N and measurable quantities. Perrin's average, $N = 7.0 \times 10^{23}$, which he gave in 1909, agreed fairly well with 6.2×10^{23}, calculated using e determined by Rutherford.

Planck's successful formula for blackbody radiation contained two constants. One the theory identified as k; the other was an ad hoc number, h, which placed a threshold on the energy ϵ that an ether mode of frequency ν could take up. Hence '$\epsilon = h\nu$', a slogan as familiar to physicists as '$E = mc^2$'. That Planck's quantum hypothesis hid a breach with classical theory did not come fully to light until Einstein and others exposed it around 1905. Two years later, Einstein brought Planck's formulation closer to the classical problem of equipartition by applying it to the vibrations (the specific heats) of elastic solids. Using Planck's radiation law, Einstein obtained specific heats that deviated from the empirical law of Dulong and Petit at low temperatures. Measurements by Walther Nernst (Berlin) confirmed Einstein's extrapolation and the insight that Planck's law applied to vibrations far removed from its original jurisdiction. The problem of the quantum moved up on the agenda of physics. At Nernst's suggestion, the Belgian industrialist Ernest Solvay sponsored a meeting in Brussels in 1911 at which all the important physicists interested in radiation and quanta came together to try to explain to one another what Planck's quantum h meant. They did not succeed.

Instances of irrationality multiplied. X-rays spread like waves but interacted with matter like particles. The same paradox appeared in the photoelectric effect, discovered by Hertz in 1887, in which ultraviolet light, which everyone knew to be a wave, knocks electrons out of metals as if (so Einstein pointed out in 1905) it was composed of particles. Radioactivity offered another instance of paradox. Thomson had made the reasonable suggestion that a gradual loss of energy by atomic electrons through radiation caused radioactivity. Disciples of Boltzmann pointed out that Rutherford and Soddy's law of radioactive decay admitted the interpretation that in a small interval of time every atom of a radioelement has the same probability of exploding. Franz Exner (Vienna) drew the surprising conclusion: the probability of radioactive disintegration and other fluctuations were chance events incalculable in principle. Many prominent physicists rejected the implied limitation on causality.

A postdoctoral student working in Rutherford's laboratory in Manchester, Niels Bohr, welcomed the impasse. His doctoral thesis had generalized the electron theory of metals, pioneered by Thomson, Lorentz, and Paul Drude (Berlin), who treated conduction electrons confined in a wire as a free gas apart from their collisions with metal molecules. Bohr came closer than they did to deriving the Wiedemann–Franz law but failed with Curie's law and heat radiation. He placed the blame on equipartition and supposed that a restriction like Planck's must be imposed on atomic electrons. That did not surprise him. His reading of Danish philosophy and literature had prepared him to expect that physical theory must occasionally hit immovable barriers.

Bohr accepted the nuclear atom and removed its defects by fiat: in their ground states, atomic electrons shall shirk their classical obligations to radiate and to perturb their common circular motion. Early in 1913, he encountered the Balmer–Rydberg formula in the form $\nu_n = K(1/2^2 - 1/n^2)$, where n, a running integer, designates a spectral line. Multiplying both sides by h and reading the result as an energy equation in Planck's style, Bohr recognized the existence of excited 'stationary states' in which electrons circulated with the immunities they had in the ground state and with energy $-hK/n^2$. Balmer lines originate in an unmotivated transition ('quantum jump') from the nth to the second stationary state. By interpreting $-hK/n^2$ as the energy of the nth stationary state and allowing ordinary physics inconsistently to rule the state, Bohr derived Rydberg's K in terms of the atomic constants ($K = 2\pi^2me^4/h^3$). This tour de force persuaded Einstein, Jeans, and other alumni of the Solvay Council that Bohr had found a way forward.

Although Bohr offered a 'correspondence principle' that hinted at how the calculations of his quantized atom could be made to agree with those of classical physics in certain limits, he emphasized that even where calculations agreed, the processes did not. His quantum jumps thus joined radioactive explosions as the first examples of natural phenomena placed formally beyond the reach of human inquiry since the time of Thales.

The profession

Most of the 700 attendees of the International Congress of Physics of 1900—about one in four of the world's physicists—held paid posts in teaching (60 per cent), industry (20 per cent), and government (20 per cent). An important novelty was the slowly growing subset of theorists. They occupied chairs primarily in Germany and Austria. In Britain and the British Empire, graduates of the Cambridge regime in mathematics held about half the permanent positions in physics in 1900. Their ability to devise and calculate the behaviour of mechanical models gave English physics an orientation that favoured the production of atomic models like those of Thomson and Rutherford.

French physics had a quite different but equally distinctive epistemology, imbued with positivism and instilled at the École polytechnique and École normale, through which over half the academic physicists active in France in 1900 passed. This positivism, with its reluctance to commit to models of the microworld, marked the approach of the Curies. No standard epistemology or hegemonic centre existed in Germany, where students customarily attended more than one university or polytechnic, or in the United States, whose professors came from a variety of universities and typically finished their training in decentralized Germany.

Most informed observers ranked Germany first among the science-producing nations around 1900 largely because of the quality of German-language publications, the preponderance of German scientific instruments, and the generous support of science institutes by several states of the Reich. Decentralization and competition were the driving principles and the guarantee that in Germany the best scientists would not be crushed, as in France, into the apex of a single educational system. Once at the Parisian apex, however, a *physicien* could regain space by accumulating posts (*cumul*). Henri Becquerel, by no means an overachiever, held three professorships in Paris, one won on his own and two passed down as if private property from his father and grandfather.

In the British Empire, little support for academic physical science came from the central government; and in the United States, none. A large source of funds for expansion in both came from individuals or corporations. Spokesmen in each country, particularly in Britain, compared this uncertain funding unfavourably with the generosity of the German *Länder*. Nonetheless, private philanthropy was expanding the material base for physics more rapidly in the Anglo-Saxon countries than in their principal rival. German statesmen of science pointed to this generosity as something to fear and emulate.

Around 1900 Britain and the United States, embarrassed to have to send their products to Germany's Physikalisch-Technische Reichsanstalt (PTR, 1887) for certification, established standards laboratories of their own. Like the PTR, the British National Physical Laboratory (1900) and the American National Bureau of Standards (1901) employed graduate physicists in increasing numbers. Naturally these recruits pressed to do research, some of which, like measurements of the blackbody spectrum at the PTR, related to fundamental problems. A few industries producing the products for testing also established research laboratories. Staff at the General Electric Research Laboratory (1900) numbered perhaps 200 in 1913, including the former academic physicists Irving Langmuir and W. D. Coolidge. A little behind General Electric came American Telephone and Telegraph (1907), Corning Glass (1908), Eastman Kodak (1912), Philips Eindhoven (1914), and Siemens & Halske (1900, 1913). (See Figure 22.)

22. The palace of electricity. The palace, a feature of the Paris Exposition of 1900, impressed visitors with its displays of colored lights. Electricity also won surprised friends by powering moving sidewalks around the extensive exhibition grounds.

Three other institutional forms created with government and/or industrial money in the early 20th century affected the tone and pace of physical science. The Nobel Prizes, endowed with the proceeds of dynamite and smokeless powder, were intended by their founder to reward those 'who, during the preceding year, shall have conferred the greatest benefit on mankind'. Although the first prize in physics, to Röntgen for X-rays, met this test, the professors soon conquered the system and rewards went more and more to academic work of no immediate practical value.

Andrew Carnegie's endowment of $10 million in 1901 for a research institute in Washington, DC, staggered contemporaries. Most of the income supported studies of meteorological subjects. To inspire similar generosity among the Reich's industrialists, a Kaiser-Wilhelm-Gesellschaft (KWG) came into being to assist in financing 'pure' research institutes. Starting from the premise that

'Science has reached a point in its scope and thrust that the state alone can no longer care for its needs,' the society's projectors pointed to the Carnegie Institution, the Nobel Institution, and a gift of a million dollars to the University of Chicago for a physics institute, as proof that Germany was falling behind. The KWG incorporated in January 1911, with a pledged private capital of around a fourth of the endowment of the Carnegie Institution. The society's institute for physics opened virtually in 1917, with Einstein as director and Planck as effective executive, but waited for a building for twenty years, and then the Rockefeller Foundation paid for it. Apparently, the society did not think that an establishment run by Einstein and Planck was likely to contribute much to German industry.

As physics professionalized, the old division of labour between the academy and the university ceased. Research increasingly became an expectation, and then an obligation, of professors. Careers depended more and more on publication in the transactions of the main scientific academies and professional societies. Among the societies, which were themselves creations of the later 19th century, those of Berlin and London, and the Institute of Electrical Engineers, were perhaps the most prestigious publishers. Among the national academies, the French, British, and American, and among regional academies those of Berlin (Prussia), Göttingen (Hanover), Leipzig (Saxony), and Munich (Bavaria) remained important outlets. In addition, general or disciplinary news journals like *Nature* (founded 1871), *La Nature* (1873), and *Physikalische Zeitschrift* (1899) disseminated research reports and good tidings.

Above the local level (universities, polytechnics, government and industrial laboratories) and the regional/national level (professional societies, academies) stood, in the scale of inclusiveness, the national associations for the advancement of science and international organizations of various degrees of specialization and permanence. The national associations collected scientists, supporters of science, and the interested public in large annual meetings that wooed inclusiveness by changing venue from year to year. The Gesellschaft deutscher Naturforscher und Ärzte was the oldest (1822), followed closely by the British Association for the Advancement of Science (1831), and, later, by similar bodies in France, the United States, and Italy.

Internationalism was as distinctive a feature of the half-century before World War I as were the nationalisms it tried to ameliorate. Encouraged by cosmopolitan or humanitarian promptings, enabled by the efficient communication and transportation system of Europe, and supported by governments seeking international agreements in technical matters, international conferences on various branches of physics tripled, from 1.7 per year around 1875 to 5.5 per year around 1905. The logical apex of the international movement in physics was an organization to coordinate the world's work in physical science in accordance with a grand research plan. In 1912, Solvay partially filled this logical space by founding an Institut international de physique. In 1914, Charles Edouard Guillaume, director of the Bureau international des poids et mesures (1875), proposed the establishment of an international association of physical societies. His timing was not good.

Chapter 6
From old world to new

Historians, philosophers, theologians, popularizers, and Nobel Prize committees concerned with 20th-century physics tend to direct their attention to TOEs, and to the microphysics of cosmogony. The emphasis privileges the mind-expanding concepts of relativity theory, quantum physics, and the systematics of elementary particles. A secondary theme stresses the control of nature by physical theory as demonstrated by the release of nuclear energy. After this demonstration, great laboratories, national and international, employing machines as expensive as battleships, multiplied 'elementary' particles as well as high-energy physicists and the army of technicians, scanners, computer specialists, engineers, grant officers, bookkeepers, secretaries, and contractors on whom TOE quests now depend. The symbol and acme of this enterprise in the short American century is Fermilab (see Figure 23).

Nevertheless, the romantic investigation of the smallest and the biggest things in the universe employs less than 20 per cent of today's physicists if the divisional breakdown of the American Physical Society adequately represents the profession at large. The majority of physicists work on condensed matter, plasmas, computers, optics, lasers, polymers, fluid dynamics, materials, atomic, molecular, and nuclear problems, questions of military interest, and border subjects overlapping chemistry, Earth

23. A small piece of big physics. A section of the Tevatron tunnel at Fermilab. Protons accelerated in an evacuated circular pipe four miles in circumference attained a velocity 99.99995 percent that of light. The pipe resides within the lower line of boxed superconducting magnets. The upper line

vsciences, biology, medicine, and engineering. They also cultivate the latest sprigs of that old trunk of *physica, meteorologica*: climatology, plate tectonics, ionospherics, oceanography, and seismology.

Legacies of World War I

While drifting westward towards the US during the 1930s, the centre of gravity of physics experienced tugs from the Soviet Union and Japan. A few numbers will illustrate the vigour of this expansion. The figure of merit is a doubling each decade between 1920 and 1940. That happened in the US for the annual harvest of PhDs in physics (50 to 200) and physicists in industry (400 to 1,600). In Japan, PhDs doubled every seven years, memberships in the Physical and Mathematical Society of Japan every eleven years (reaching 1,100 in 1940), and research laboratories every twenty years (to 325). The rapid growth of physics in the Soviet Union may be indicated by an eight-fold increase in universities and colleges, a six-fold increase in scientific staff, and a twenty-fold increase in employees of the Academy of Sciences.

The feedback mechanism for this explosive growth, and for slower increases in European centres, was the discovery of the importance of physics for national defence, industrial development, and what Italian fascists and Japanese militarists called 'autarky'. The Great War had shown that determination and technology could make a nation independent of others for products and raw ingredients. Thus it was with nitrogenous fertilizers in Germany and with pharmaceuticals, fine chemicals, electrical appliances, and optical glass in the Entente powers. Among the first research establishments set up in Japan and the Soviet Union were optical institutes charged with perfecting glass for military purposes. In the late 1920s, Italy followed suit with an institute for optics, one of the few successful state-supported high-tech research endeavours during Mussolini's regime.

Two other industries virtually created during the war, aeronautics and radio, gave employment to physicists in all developed countries after it. Government and industrial laboratories for radio and telephony multiplied, perfecting vacuum tubes, receivers, relays, and transmitters, promoting miniaturization and developing sound recording and playback. Physicists so employed helped to engineer several simultaneous revolutions by providing hardware for large-scale research, radios for domestic leisure, and the loudspeaker for politicians and propagandists. Fascists and Nazis battened on the big lie, the mass rally, and the amplifier.

The science and rhetoric of World War I still confront us in weather reports, which record the vision of Vilhelm Bjerknes (Bergen) and his school of climatologists. Cut off from meteorological information from the West during the war, they filled the hole with the theory of fronts between struggling air masses. They chose their terminology with an eye to trench warfare. Massive data about the upper atmosphere, collected to guide the flights of planes and shells, stood ready to test and extend their theories after the war.

In addition to its legacy of hardware and practically minded scientists, World War I confronted demobilizing physicists with discoveries made in sheltered niches during it. Einstein had pondered the curious fact that a body's resistance to motion has the same measure as its acceleration under gravity. It followed that all observers can consider themselves at rest in their reference frames if they attribute any acceleration they feel to gravity. This equivalence permitted Einstein to geometrize gravitation 300 years after Galileo had intuited the possibility. The trick was to use non-Euclidean geometry. A wheel spinning at relativistic velocities has a different geometry than when resting because the circumference suffers a Lorentz contraction and the radius does not. Owing to the equivalence of acceleration (in this case of circular motion) and gravity, Einstein inferred that massive gravitating bodies warp the space around them so as to capture other bodies. Depending on the velocity and direction with which a celestial object approaches the warp, it can be swallowed like an apple, acquired as a planet, or deflected as a comet. Just like a Cartesian solar vortex! Einstein completed his 'general theory of relativity' in November 1915, the year in which his friend and colleague Fritz Haber enriched the art of war with poison gas.

According to general relativity, light should swerve when passing through warped space. Einstein suggested testing the prediction by comparing the direction of stellar rays that grazed the Sun during a total solar eclipse with their direction six months earlier. In 1918, American astronomers tried to detect the expected shift. They failed. A year later an English expedition found it. The confirmation aroused immense public interest in a world eager to raise its gaze to the stars.

Bohr's quantum theory of the atom prospered mightily during the war. In neutral Denmark, he and his Dutch student, Hendrik Kramers, developed the correspondence principle, and in neutral Sweden Manne Siegbahn (Stockholm) and his students made precision measurements of high-frequency spectra that gave access to the deep structure of atoms. These measurements exploited two capital prewar discoveries. In 1911–12, Max von Laue (Munich), William Lawrence Bragg (Cambridge), and William Henry Bragg (Leeds) showed that X-rays can be refracted and reflected by crystals, which made possible X-ray spectroscopy and tipped the wave–particle duality towards waves; and in 1913–14 Henry Moseley (Manchester, then Oxford) inaugurated X-ray spectroscopy by detecting high-frequency line spectra characteristic of the elements from which they came. Moseley found that the strongest line from an element Z had a frequency that obeyed the simple Balmer-like formula, $\nu=K(Z-1)^2 \times (1-1/2^2)$. He interpreted the formula as a confirmation of the concept of atomic number and a signpost to the innermost region of the atom.

Arnold Sommerfeld (Munich), who was too old to fight and impractical in war work, developed Bohr's theory by introducing additional quantum numbers. With them he accounted for X-ray and complex optical spectra. His systematic presentation of his postulates and results, *Atombau und Spektrallinien* (1919), gave demobilizing physicists the score of what he called the 'atomic music of the spheres'. Sommerfeld's music, though magnificent, was too formal for Bohr's taste as it introduced quantum numbers and ad hoc limitations on electronic transitions without much physical justification. Bohr preferred his method of intuition and conjecture, which only he, following what Einstein admired as his 'unfailing tact', knew how to use.

The third fundamental wartime novelty in basic physics obtruded when Rutherford, snatching time from his defence work, inquired how close to the nucleus Coulomb's law held. He found a few deviations of a peculiar kind in encounters between hard alpha rays and air molecules: instead of being repelled, the alpha disappeared and a faster, apparently lighter particle emerged. Rutherford supposed that he witnessed a game of marbles in which the alpha entered the nuclear precinct and knocked out a hydrogen nucleus (a 'proton'), and conjectured that nuclei consisted of alpha particles, protons, and neutral particles each made up of a beta ray and a proton.

From electrons to stars

In the summer of 1922, Bohr gave inspiring, perplexing lectures at Göttingen in which he claimed that the correspondence principle determined how many elements fell into each row of the periodic table. He made do with two quantum numbers. Among his auditors was Sommerfeld's prize student Wolfgang Pauli, who three years later assigned two more quantum numbers to each electron to derive the lengths of the rows of the periodic table, 2, 8, 18, 32. He required the 'exclusion principle' that no two atomic electrons can have the same values of all four quantum numbers.

Its implication that antipathy reigned among electrons showed that some concepts of Greek *physica* remained useful; and its rapid acceptance measured how far atomic physicists had accustomed themselves to the irrationality of the microworld.

In the same year, 1925, another product of Bohr–Sommerfeld rearing, Werner Heisenberg, fresh from a stay with the strict mathematical theorist Max Born (Göttingen), found a calculus relating only observable features of spectral lines. The calculus presented frequencies and intensities in square arrays ('matrices') whose entries (the intensities) were located by quantum numbers defining possible electronic transitions. Heisenberg, his former fellow student Pascual Jordan, and their teacher the formalist Born quickly sophisticated Heisenberg's insight so effectively that very few physicists could understand it. Fortunately for their self-esteem and sanity, Erwin Schrödinger (Zurich), a descendant of Boltzmann's school, found another calculus without recourse to matrices. Rather, he picked up a hint from the doctoral thesis of Louis de Broglie (Paris), who used relativity theory idiosyncratically to associate a standing wave with an electron moving in a stationary state. Schrödinger devised an equation for this wave (subsequently 'the ψ-wave' or 'wave function') that depended on the electron's energy. Calculators soon discovered that they could obtain the numbers in Heisenberg's matrices by using Schrödinger's waves.

The almost simultaneous discoveries of wave and matrix mechanics followed two different paths from the same demonstration: that an X-ray can strike an electron much as one billiard ball does another. This 'Compton effect' (1922, after Arthur Holly Compton, St Louis) stampeded physicists into taking seriously the concept of the light particle ('photon') that Einstein had suggested as a 'heuristic hypothesis' in 1905 to account for the photoelectric effect. De Broglie's coupling of a wave to a particle accepted the photon, whereas Bohr, who had rejected light quanta, tried to elude Compton's billiard-ball theory. His failed effort helped Heisenberg invent quantum mechanics. After experiments decided for Compton, Bohr gave equal weight to wave and particle properties. Bohr added the qualification that no realizable experiment can require a description employing these contradictory characteristics simultaneously ('complementarity', 1927).

Schrödinger tried to interpret the ψ-wave realistically as a measure of the electrical density at every point in the atom. He did not succeed. Physicists reluctantly accepted Born's elucidation that ψ^2 at any place gives the probability of finding the entire electron there. Heisenberg then drew from the mathematical formalism the conclusion that the microphysicist can never do better than reckon the probable, that not even a Demiurge could determine the exact position and corresponding momentum of an electron ('the uncertainty principle', 1927). To Einstein and other senior physicists this conclusion was a derogation of duty. Several times Einstein constructed thought experiments that garnered more information than quantum mechanics allowed, and every time Bohr escaped by finding an inconsistency or omission in Einstein's examples. The defeat of his last attempt, made with two colleagues in 1935, persuaded almost all physicists who still took an interest in the argument that quantum mechanics does give a complete though probabilistic atomic and molecular physics.

By 1935, the quantum mechanics of electrons and radiation rested on a relativistic theory invented by Paul Dirac (Cambridge, 1928). Its results had been as surprising and gratifying as the radio waves unexpectedly generated by Maxwell's equations. Imposing the requirement of relativistic invariance made the electron's wave function describe particles with a fourth degree of freedom and either positive or negative energy. Dirac interpreted the extra degree as an intrinsic (but non-mechanical) spin, thus rationalizing Pauli's quantum numbers and the concept of electron spin proposed to improve spectral systematics by Samuel Goudsmit and George Uhlenbeck (Leyden) in 1925. The negative-energy solutions appeared to describe a particle with positive charge. One that fitted the description (the 'positron') quickly turned up in a context soon to be described. Meanwhile, several researchers, including J. J. Thomson's son George Paget Thomson (Aberdeen), detected the 'de Broglie wave' of electrons. The subsequent award of a Nobel Prize to Thomson for showing that the electron is wave-like thus oddly complemented the prize his father had won for showing it to be particle-like.

Dirac's relativistic theory supposed that a high-energy photon could create an electron–positron pair, which, by subsequent mutual annihilation, could make another photon. Incorporating this demiurgic idea, theorists developed quantum electrodynamics, an account of electrons and photons real and virtual that predicted values of many measurable parameters to exquisite accuracy once they had developed rules to remove the infinities that pestered their calculations. That did not happen until after World War II.

The meagre flux of high-energy alpha particles from natural radioactive sources did not score enough hits to explore nuclear architecture effectively. Several physicists proposed to increase the flux by accelerating beams of protons to the many hundreds of keV (1,000 electron volts) deemed necessary to jam one into a light nucleus. Here the subtleties of quantum mechanics, brought to Rutherford's laboratory at Cambridge by the imaginative Russian physicist and humourist George Gamow, came to the aid of the marble shooters.

Gamow had noticed that the probabilistic ψ-calculus predicted that a proton could sometimes enter ('tunnel into') a nucleus with less energy than it would require if it behaved like a marble and the nucleus like a brick wall. Rutherford's lieutenant, John Cockcroft, in collaboration with Ernest Walton and the major electrical manufacturer Metropolitan Vickers, exploited tunnelling to reduce lithium atoms to alpha particles with 300 kV protons.

By 1932 they had taught their apparatus to hold 700 keV and disintegrated lithium in quantity.

Meanwhile, Ernest Lawrence (Berkeley) had devised a way to add energy in small steps to a spiralling charged particle so that the entire accelerating potential never had to be maintained on any part of the apparatus. The key to his 'cyclotron' is that at nonrelativistic speeds the frequency with which a charged particle circulates when confined by a magnetic field does not depend on its energy. Lawrence and his students placed two hollow D-shaped electrodes within a squat cylinder with a proton source at its centre, exhausted the space, applied an oscillating electric field across the gap between the 'dees', slathered on sealing wax, shimmed the magnet, and, at the end of 1932, had a tiny beam at 1 MeV (million eV) (see Figure 24). That same year James

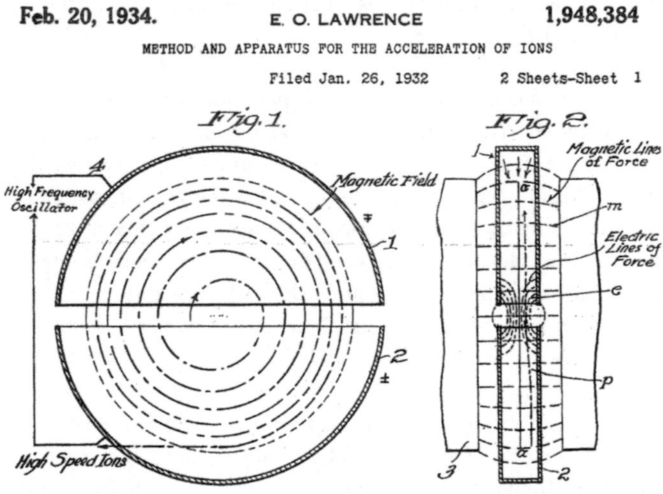

Feb. 20, 1934. E. O. LAWRENCE 1,948,384

METHOD AND APPARATUS FOR THE ACCELERATION OF IONS

Filed Jan. 26, 1932 2 Sheets-Sheet 1

Fig. 1. Fig. 2.

High Frequency Oscillator · Magnetic Field · High Speed Ions · Magnetic Lines of Force · Electric Lines of Force

24. Patent on the cyclotron. The cyclotron principle—that the accelerated particles' angular frequency, $v/r = (e/m)H$, is independent of its energy at nonrelativistic speeds—allows the spiral motion to take place entirely under a constant magnetic field H. As shown on the right, the electric and magnetic fields automatically focus the particle beam.

Chadwick (Cambridge) detected Rutherford's nuclear neutral particle ('neutron'). The simplest combination containing it, the deuteron, the nucleus of heavy hydrogen ('deuterium'), was isolated, again in 1932, a wonder-filled year for physicists if for no one else at the depth of the Great Depression.

As available energies increased, so did the number of elements transformed or disintegrated. The marble model persisted, however, until 1934, when Marie Curie's daughter Irène and her husband Frédéric Joliot (Paris) observed a delay between absorption and admission. Soon nuclear physicists and chemists were making 'activities' throughout the periodic table by shooting protons, deuterons, alpha particles, and neutrons at every element they could procure. By 1939, the ever-enlarging Berkeley cyclotron was producing prodigious currents of 16 MeV deuterons. It often laboured around the clock to make radioisotopes with desirable properties as tracers (notably carbon-14) and therapeutic agents (phosphorus-32). The cyclotron was preeminently and, with its large attendant staff, capital costs, and operating expenses, symbolically American. Only eight existed outside the US at the outbreak of World War II. A man from Berkeley had to be summoned to get several of them started.

Just as the marble analogy blinded physicists to artificial activities, the expectation that the products of a transformed element would lie close to it in the periodic table blinded them to nuclear fission. Enrico Fermi and his group (Rome) saw and missed it in the course of irradiating uranium with neutrons. Lise Meitner and her chemist colleagues Otto Hahn and Fritz Strassmann (Berlin) persisted for years in identifying the products of uranium irradiation as 'transuranic'. Eventually, the chemical evidence persuaded Meitner's colleagues that the uranium nucleus could be split into two elements of moderate atomic weight. Meitner, then an émigrée in Siegbahn's laboratory, and her nephew Otto Frisch, then working in Bohr's institute, proposed the mechanism of fission, which Bohr clarified and broadcast to the world in 1939. Everyone knowledgeable realized that the neutrons released by fission in a large sample of uranium might provoke a chain reaction.

After the excitement of 1920, interest in Einstein's stable, almost flat, closed universe flagged for a decade. However, his equations also allowed expanding, contracting, and oscillating universes. The first to treat expansion as the true state of affairs was Georges Lemaître, a Belgian priest who had studied at Cambridge and Harvard, where he learned about Edwin Hubble's (Caltech) capital discovery that other 'island universes' (galaxies) are receding from ours with speeds proportional to their distances from us. In 1931, Lemaître proposed that the expansion spreading the galaxies began when the universe, encapsulated in a single atom, exploded.

Just after the explosion, everything was radiation and maybe electrons. Where then did Thales' water come from? The question of the origin of the elements drew George Gamow, his Hungarian colleague Edward Teller (Washington, DC), and the world's leading nuclear theorist Hans Bethe (Cornell), all immigrants to the US, into the expanding universe. Gamow supposed that stellar interiors provided the ovens that cooked protons into the elements and that his tunnelling mechanism kept down the temperature required. In 1938, Bethe succeeded in defining the nuclear syntheses that power the stars, which did not require or reach a temperature at which nuclei heavier than helium's could be made. Gamow therefore relocated his ovens (unnecessarily, as it turned out) to the hotter regime of the early expanding universe.

In 1942, in the last of the Washington meetings organized by Gamow and Teller that had helped set the agenda of American theoretical physics in the 1930s, the participants agreed that 'the elements originated in a process of explosive character, which took place "at the beginning of time" and resulted in the present expansion of the universe'. Thus the physics of the laboratory, where the energies required to add a nucleon to a nucleus could be measured, made contact with speculations about the state of the universe at grossly inaccessible times and places. At this point Gamow broke off speculating about the great explosion at the origin of time to help the US navy improve its gunnery in the here and now.

Hotness and coldness

The discovery of radioactivity breathed new life into the subject of atmospheric electricity created by Franklin's theory of lightning. Radioactive substances near the Earth's surface give off radiation that ionizes the lower atmosphere and keeps Earth from cooling at the rate Lord Kelvin had calculated. As expected, the ionizing radiation declined with height above the Earth's surface. Consequently the discovery that at higher altitudes ionization increases bewildered Victor Hess (Vienna, 1912) and other daredevil physicists who measured radiation from balloons. Aristotle's problem arose again: was the cause located beneath or above Moon? Hess judged it to lie outside the solar system since it operated with equal strength day and night. After World War I, Millikan took up the question using war-surplus sounding balloons. In his opinion, incoming 'cosmic rays' were photons created in space during the amalgamation of four protons. Absorption measurements and calculations of the energy released (via $E = mc^2$) at first favoured Millikan's fantasy that cosmic rays were the 'birth cries' of atoms. In 1929, Walther Bothe and Werner Kohlhörster (Berlin) found that most of the charged rays at sea level could penetrate 10 inches of gold and so could not be photons.

The Earth itself provided the instrument needed to decide whether cosmic rays are charged particles. Bruno Rossi (Florence), Lemaître, and Manuel Vallarta (MIT)—adapting Carl Størmer's (Oslo) calculations of the paths of particles causing the aurora borealis—showed that the Earth's magnetic field would bend the trajectories of incoming charged particles by an amount dependent on latitude. An international team led by Compton and deploying standardized ponderous electroscopes supplied by the Carnegie Institution's Department of Terrestrial Magnetism, confirmed that the rays obeyed the calculations.

The latitude survey counted all the cosmic rays that penetrated its instruments. Bothe and Kohlhörster had followed the path of one and the same ray through their gold brick. They had considered only events in which Geiger counters placed above and below the brick fired together. Their method excited Rossi and his fellow student in Florence, Giuseppe Occhialini, to do better. Exploiting vacuum tubes available for radios, they detected counter coincidences electronically. Thus they discovered that a cosmic ray can sail through a metre or more of lead, implying energies of several GeV (1,000 MeV).

Electronic counters when arranged to trigger a cloud chamber made possible efficient self-portraiture of a cosmic ray. This chamber, invented before the war by C. T. R. Wilson (Cambridge) and improved after it by P. M. S. Blackett (also Cambridge), recorded tracks by condensing water droplets around ions created by the passage of energetic charged particles. The intensity of ionization and the curvature of the tracks in a magnetic field gave information about the particles' charge, mass, momentum, and energy. Blackett's group inspected 400,000 tracks in 1924 to find six showing Rutherford's nuclear marble game. By hooking up Rossi's coincidence circuitry to the cloud chamber, however, they quickly obtained many fine pictures of the tracks of positive electrons. That was shortly after Carl Anderson (Caltech) chanced to find a beautiful example on one of the 3,000 exposures he made without coincidence circuitry.

The positron put in its appearance in 1932, the year of the neutron, deuteron, cyclotron, and Cockcroft–Walton. Five years later, Anderson and others at Caltech, Harvard, and Tokyo found the tracks of what they called 'mesotrons' or 'mesons'—particles weighing around 200 electron masses. The weight agreed with that of the hypothetical 'Yukon', which Hideki Yukawa, the first important domestically trained Japanese theorist, had proposed in 1935 as the mediator of the 'strong force' between neutrons and protons. Physicists had already accepted a 'weak force' to account for beta decay and an imponderable uncharged 'neutrino' to conserve energy in the process. Naturally, they identified Anderson's mesotron with the Yukon.

In 1943, Rossi, resettled in the US, determined the mesotron's half-life at rest by measuring the delay between its arrival on Earth and the emission of the electron into which it decays. The result, two microseconds, was not long enough to bring mesotrons down through the atmosphere in the numbers observed at the surface. Since, however, they move at relativistic speeds, they live detectably longer in motion than at rest, and so delivered the first direct confirmation of Einstein's time dilation.

Physicists gained the keys to another new kingdom at the same time they discovered cosmic rays. In 1911, Heike Kamerlingh Onnes (Leyden) and his associates found that the electrical resistance of pure metals plummeted suddenly to zero at liquid-helium temperatures. This phenomenon, to which he gave the name 'superconductivity', was not his greatest surprise in the kingdom of the cold. Liquid helium had an anomalously low density and would not freeze. These hints stimulated research after the war following a pattern similar to that of nuclear physics. A new-world laboratory challenging the old leader invented new costly apparatus that magnified effects a hundredfold. Stranger phenomena soon came to light. Theorists competed to explain them by applying quantum mechanics to fields founded in the decade before World War I.

The new-world site that challenged the old (in this case Leyden) was again Berkeley. The counterpart to Lawrence, William Francis Giauque, attained temperatures within a degree of absolute zero in 1933 by applying 'adiabatic demagnetization'. This technique cools a paramagnetic crystal to liquid-helium temperatures where the little thermal energy it retains resides in the vibrations of its lattice and the disordered spins of its atoms. Giauque used a strong magnetic field to line up the spins and liquid helium to carry away the heat thus generated, as in the isothermal compression of a steam engine. Having insulated the specimen thermally, he turned off the field, the spins regained their disorder at the expense of the lattice vibrations, and the temperature fell, as in an engine's adiabatic expansion, to half a degree absolute. By 1938, he had arrived within 0.004° centigrade of absolute zero. Experimentalists thus acquired a tool for studying superconductivity and other odd phenomena in regions well below the liquefaction point of helium. The odd phenomena, which included a sudden, million-fold jump in heat conductivity and an abrupt vanishing of viscosity in helium, gave rise to the concept of superfluidity (1938).

Effective theories of superconductivity and superfluidity drew on a quantum reformulation of the old electron theory of metals synthesized by Sommerfeld and Bethe in 1933. Building on the insights of Heisenberg's student Felix Bloch (Leipzig, then Stanford), they treated unbound electrons in crystals at low temperatures as free except for the exclusion principle, which forces them to fill up the lowest quantum states available. Only those near the top of this electron sea can take up additional energy as the temperature rises and so sail freely along or above it. The new model agreed with the old theory where it had worked, and explained its failures (as in specific heats) where it had not. These results, obtained by European theorists, pertained to ideal solids. Beginning in 1933, Eugene Wigner and his student Frederick Seitz (Princeton) began the hard calculations of what Pauli called 'dirty effects' (caused by impurities), and other American groups, notably John Slater's at MIT, soon mucked in.

The free electron model did not account for superconductivity. That became more embarrassing as information about it mounted during the 1930s. There were two main centres of research: Oxford, where Franz Simon, Fritz London, and other German-Jewish refugee physicists, descended intellectually from Nernst and, in Simon's case, Giauque, set up a thriving low-temperature laboratory with the help of British industry; and Moscow, where Peter Kapitza, a former associate of Rutherford, and the theorist Lev Landau, worked in the institute given to Kapitza as consolation for his unwilling separation from Cambridge. Although some of these men glimpsed the nature of a workable theory of superconductivity, they kindly allowed Americans to invent it.

Americanization

During World War II, physical science and engineering created radar, which led to microwave technologies and the laser; atomic bombs, which brought nuclear power; the V-2 rocket, which launched the aerospace industry; and the 'bombe', the essential device for breaking the Enigma codes, which, with electronic analysers constructed for calculating ballistic trajectories, pioneered the electronic computer. This intrusive device emerged in the late 1940s in materializations of Alan Turing's conception of a 'universal machine', which related to the computer as Carnot's theory did to the steam engine. The ENIAC, the first postwar electronic computer, was rich in vacuum tubes (some 17,468), poor in memory, and programmable only by changing its wiring. Developed at the University of Pennsylvania under an army contract to calculate firing tables, it soon switched to working on the hydrogen bomb. Its successor, the MANIAC, built on the architecture created primarily by John von Neumann (Princeton), contained its own instructions and could be reprogrammed without rewiring. Von Neumann architecture has survived throughout the revolutionary advances made possible by transistors, microprocessors, and gigabyte memories.

As the only major belligerent to emerge from the war rich and relatively unscathed, the US was best positioned to capitalize on the computer and other inventions. To augment its head start in exploiting allied wartime inventions, the US sent scavengers behind its advancing armies to seize information about developments in Germany. In the transfer of Werner von Braun's rockets and technicians to Alabama ('Operation Paperclip'), the US picked up men as well as hardware without inquiring zealously about their degrees of Nazism. The French, British, and Russians also took what they could.

Americanization meant not only an emphasis on big machines and large collaborations, and hurried competition fostered by ambition and federal funding agencies, but also a neglect of matters that the generation of Bohr and Einstein had thought important. Just as Newton's followers in the 18th century dropped his scruples about distance forces and ignored his rules of philosophizing, so postwar Americans did not bother themselves with the enigmas of quantum physics or, in general, with the high culture of the prewar Europeans. The uncomplicated unlearnedness of Lawrence was closer to the norm than the tortured culture of Robert Oppenheimer.

The Americanization of theoretical physics made its postwar debut in a conference held in June 1947 at Shelter Island to insulate the participants from the demands society had placed on 'brilliant young' physicists—the only type, apart from Einstein, it knew. The planners consulted Pauli, then at Princeton, and ignored his suggestion to reconvene the creators of quantum mechanics. They opted for a small specialist meeting featuring upcoming Americans, homegrown theorists from Oppenheimer's school at Berkeley and Caltech and other youths whose science the war had honed. The plan of the meeting, drawn up by John Wheeler (Princeton), reads like the agenda for a fast-paced attack on a wartime problem. The brilliant young things were to begin with the question whether a logical, consistent, and comprehensive quantum theory exists; if yes, to frame it; if no, to define the limitations of existing theory and propose ways to extend it.

Two-thirds of the participants were American by birth and most of the others were émigrés of the 1930s. Bohr's former collaborator Kramers (Leyden) urged in vain for consideration of Copenhagen's old-world view that the troubles of quantum electrodynamics ('QED') could not be overcome without a better classical electron theory on which to apply the correspondence principle. The intellectual leader of the meeting was Oppenheimer. Performing as he had when director of Los Alamos, he went straight to current problems like the demoralizing infinities of QED and the puzzling mesotron that lived too long to be a Yukon. Robert Marshak (Rochester) made the generous suggestion, soon confirmed, that the Yukon was not the only meson in the world. As for the infinities, pragmatic 'renormalization' of QED by Richard Feynman and Julian Schwinger (and independently by Yukawa's classmate Sin-Itiro Tomanago) saved the phenomena to as many decimal places as anyone could measure.

A second conference, held in 1948 with most of the same participants but with Bohr in place of Kramers, inspired an instructive clash between new and old. Feynman (Cornell) sketched his version of QED with diagrams of the sort that soon became indispensable. Bohr did not like them: they violated the uncertainty principle by assigning definite paths to particles and they represented positrons incomprehensibly as electrons moving backward in time. No matter. Feynman and Schwinger (Harvard) were the new world-class theorists.

The US had learned to produce not only theorists of the Jewish-cosmopolitan-extrovert type like Oppenheimer, Feynman, and Schwinger, but also of an introverted, unassuming, gentile type. John Bardeen began neither as a physicist nor as a New Yorker, but as a petroleum engineer in the Midwest. Sickening of oil, he enrolled in graduate school in Princeton and worked on solids under Wigner. He started as a physicist at Bell Labs where, at the end of the war, he joined a small group under William Shockley (a student of Slater's from MIT) dedicated to making an amplifier using the semi-conductors silicon and germanium, known intimately for their use in radar detectors. Bardeen's mastery of the quantum theory of solids together with the experimental inventiveness of Walter Brattain and the support of Shockley resulted, in 1947, in the invention of the transistor and a revolution in electronics. While Bardeen was in Stockholm in 1956 to collect the Nobel Prize awarded to him, Brattain, and Shockley, his theorist colleagues Robert Schrieffer and Leon Cooper (all then at the University of Illinois) continued work on his idea that interactions between vibrations of the crystal lattice and electrons skimming the surface of Sommerfeld's electronic sea supplied the mechanism of superconductivity. It was back to Stockholm for Bardeen, the only two-time winner of the Nobel Prize in Physics.

New business

In 1944, drawing up plans for the postwar era, Lawrence reckoned that his laboratory would require around $85,000 annually and gifts in kind from the army's Manhattan Engineer District to do well in peacetime. A year later, knowing that the bomb project would succeed, he raised his expectations a hundredfold. In the last days before the District became the civilian Atomic Energy Commission (AEC, 1947), it gave Lawrence hundreds of kilo-dollars' worth of electronics for incorporation into the three major accelerators his laboratory had in hand. One of these, the 184-inch synchrocyclotron, started up just before the AEC did, with the presumed capacity of making mesons. Attempts to capture them on photographic emulsions failed, even after Cecil Powell (Bristol) and his cosmic-ray group had confirmed Marshak's conjecture by photographing the birth of a Yukon (see Figure 25). Its parent, the 'pi-meson' or 'pion', is the mediator of the strong force. Giulio Lattes (Brazil), who had worked with Powell, came to Berkeley and showed the cyclotroneers how to photograph pions. As in induced radioactivity and nuclear fission, a Berkeley accelerator had produced the means but not the men to make a discovery realized in Europe with much simpler equipment. But the future belonged to the big machine.

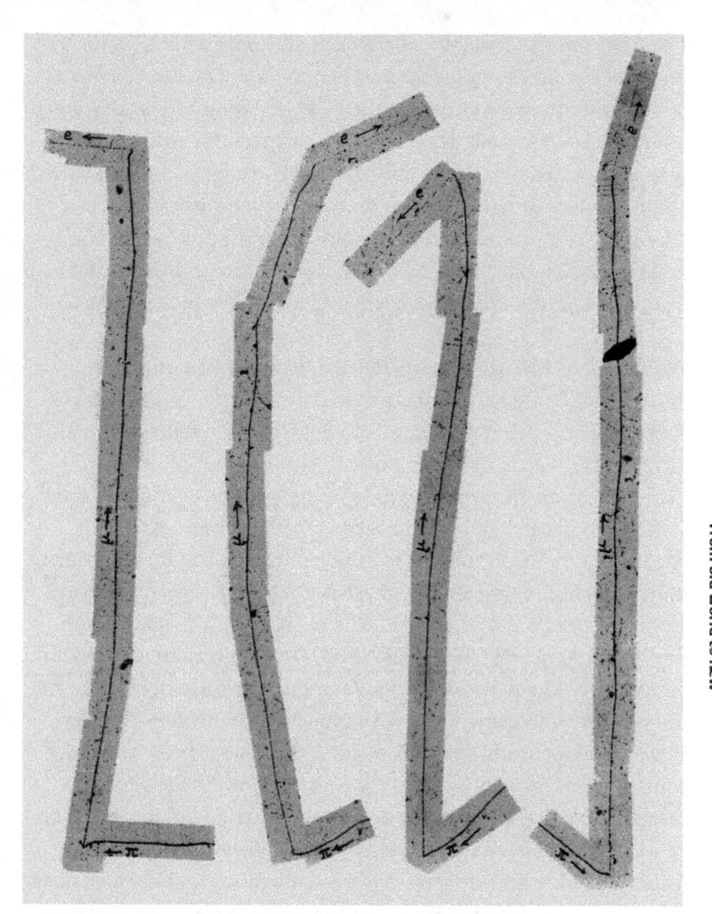

25. Birth and death in the microworld. An emulsion stack exposed to cosmic rays gave the first proof of the two-meson hypothesis. In the example furthest left, a pion on the right strikes a molecule, giving rise to a muon that proceeds upwards until it decays into an electron and a neutrino.

The AEC, uncertain whether better acquaintance with the pion might not be necessary for the next generation of nuclear bombs or power, authorized a machine in the 2–3 GeV range for its laboratory at Brookhaven on Long Island. Berkeley kept ahead with a 6 GeV version expected to produce antiprotons. The huge energies involved required a new design, in which protons could be accelerated to relativistic speeds not by spiralling out under the pole pieces of a large magnet, but by being confined by small fast-acting magnets in a narrow pipe laid out in a circle over 36 metres in diameter.

When completed in 1954, the 'Bevatron' was an engineering marvel that could accelerate protons through four million turns (300,000 miles) in 1.85 seconds to 6.2 GeV. The following year, a group led by Fermi's former student Emilio Segrè (Berkeley) detected the antiproton. The detection was noteworthy not only for confirming the general theory of antimatter but also for revealing serious difficulties in apportioning credit for discoveries in big-machine physics. The physicist-builders of the Bevatron certainly deserved some credit. The designer of the quadrupole focusing magnets Segrè's group used, Oreste Piccione, an Italian émigré working at Brookhaven, demanded it. But only Segrè and his senior collaborator Owen Chamberlain received the Nobel Prize. Piccione's subsequent suit for a share was thrown out of court not on the merits, but because he had waited too long to bring it. Hundreds of physicists, from graduate students through veteran scientists, may now work on a single 'experiment'. The feeling of being another cog in a great machine, and the certainty that the project leaders would get the credit for its results, drove many individualists into other sorts of work.

Nonetheless, accelerator laboratories multiplied. While the Bevatron settled down to making strange particles and its attendants developed the bubble chamber—a detector that rivalled the main machine in cost and complexity and depended on cryogenics developed for the hydrogen bomb—competitors first nationally and then internationally came to threaten

Berkeley's lead in high-energy physics (HEP). Fermilab, founded near Chicago in 1967, took the lead. Berkeley compensated by diversifying into projects of more immediate public utility and by cooperating more fully with other laboratories. In this way it copied Europeans who, being unable after the war to match US investments in HEP, banded together to form the Centre Européen pour la Recherche Nucléaire (CERN), which now has the largest accelerator in the world. Its circumference of 27 km crisscrosses the French-Swiss border near Geneva (see Figure 26).

While pioneering in international scientific collaboration, CERN has been a cornucopia of elementary particles. Most recently, in 2012, it had the highly hyped honour of detecting the heavy particle postulated by Peter Higgs (Edinburgh) in the 1960s to

26. Very big science. A view of CERN indicating the route of the Large Hadron Collider under France and Switzerland. In 2009 its greater capabilities caused the Tevatron to shut down. Other acronyms (ALICE, ATLAS) designate experimental sites; SPS (Super Proton Synchrotron), CERN's first big instrument, now supplies the main ring with its initial beam.

account for the mass of other particles. In the interim, physicists reduced nucleons and other heavy particles to combinations of 'quarks' held so firmly together by 'gluons' that experiment can never pry them apart. According to this standard model, whose playful American terminology jars pleasantly with Whewell's classical coinage, the nuclear, weak, and electromagnetic forces were one and the same during the very early universe.

Immediately after World War II, English scientists who had worked on radar copied Galileo and directed their war-surplus equipment towards the heavens. The first important groups grew in Cambridge (J. A. Ratcliffe, Martin Ryle) and Manchester–Jodrell Bank (Bernard Lovell). They mapped radio sources in the solar system and identified many in other galaxies. In one source, Maarten Schmidt (Caltech) found evidence of a red shift implying great distance, which, with its observed brightness, indicated an output of energy some hundreds of times that of our entire galaxy. Many such 'quasi-stellar objects' (quasars) then put in an appearance, as did 'pulsars' (Jocelyn Bell et al., 1967), which emit staggering bursts of energy at regular short intervals. These objects, and other odd stellar species uncovered by X-ray, infrared, radio, and visual observations, helped rekindle interest in cosmology.

When the US entered radio astronomy around 1950, it possessed the first large telescope built to detect centimetric microwaves. In this region, in 1965, two physicists at Bell Telephone Laboratories discovered an annoying universal background radiation while surveying sources of radio noise. With the help of Robert Dicke and others at Princeton who unknowingly had revived Gamow's cosmological ideas, the Bell discoverers, Arno Penzias and Robert Wilson, identified the ubiquitous noise with radiation left over from the aboriginal explosion (or 'Big Bang') of the universe. According to Gamow and his student Ralph Alpher, radiation had gone its own way, disjoined from matter, shortly after the Bang and subsequently cooled as it expanded with the universe. The noise that bothered Bell Telephone had a temperature not far from that expected from the expansion rate and Planck's formula. The community of cosmologists was then enjoying a dispute between Big Bangers (Gamow and the Princeton group), and Steady-State Men, who supposed that spontaneous creation of hydrogen atoms compensated for the dilution effected by expansion (Fred Hoyle, Hermann Bondi, and Thomas Gold, all Cambridge). The background radiation settled the question.

Radar expertise gave birth to a discovery of greater importance to most people than cosmic noise. In 1954, Charles Townes, who worked at Bell Labs from 1939 to 1947 on centimetric radar, combined war surplus with Einstein's old idea of 'stimulated emission', a quantum process by which incoming light entices an atom in an excited state to radiate. The resulting device, the maser, which two Nobel laureates at the university (Columbia) to which Townes had moved from Bell Labs told him would not work, worked on molecules. Townes and others soon found ways to promote the maser to optical frequencies. The uses of the new device, the laser, in physics and everyday life are legion: holography, rock concerts, surgery, light art, CD players, barcode readers, and so on. The *Handbook of Laser Technology* published fifty years after Townes' invention of the maser required 2,665 pages to describe them all.

Steering missiles, rockets, space probes, and nuclear submarines, testing nuclear explosives above and under ground and water, monitoring test-ban treaties, and searching for uranium and other strategic materials required exact information about Earth, ocean, and atmosphere. The relevant specialties—seismology, ionospherics, meteorology, and oceanography—had received little funding or prestige between the wars in comparison with nuclear physics. After World War II the US government threw money at them and joined enthusiastically in underwriting the multi-billion-dollar International Geophysical Year dedicated to the worldwide collection of pertinent data. At the same time, 1957, Earth sciences received a pleasant jolt from the launch of Sputnik. The National Aeronautics and Space Administration (NASA), set up to close the 'space gap' and to put a man on the Moon, needed to know what its rockets, satellites, and space vehicles would encounter en route, in orbit, and on arrival. Bits of moon rock, displayed around the US, promoted Earth scientists as the mushroom cloud had nuclear physicists.

Flights of meteors, the solar wind, and the Earth's magnetosphere have become as predictable as the tides and even the weather, which is now usually forecast more accurately by meteorologists than by astrologers. This improvement was another dividend from military investments. Before World War II, many meteorologists rated the concept of fronts as 'risky' and the prospect of a dynamical weather theory as 'unpromising'. After the war, Jule Charney (Princeton, MIT) and von Neumann adapted geostrophic models (winds blowing along isobars) to computation by electronic computers. Models incorporating barocline instabilities (mismatches between the gradients of temperature and pressure) then developed together with computer capability and data collection via radar and satellite. These conceptual and computational advances have helped to identify anthropogenic contributions to climate change.

The stimulation of interest in Earth sciences by government investment after World War II led to a fundamental discovery. Oceanographers detected that seafloor sediments and high ridges lying in the middle of the great oceans were relatively youthful, and that the foothills of the immense mid-ocean mountain chains enjoyed a surprisingly benign temperature. Further investigation disclosed that the temperate regime arose from melted rocks bubbling up from cracks in the ocean floor. In 1960, Henry Hess (Princeton) conjectured that the warmth and youth of the sediments indicated that molten magma rose from the Earth's interior at these ridges and spread the seafloor apart. That could not continue for very long unless the spreading floor disappeared somewhere. Obvious loci for such brutal events are fault lines showing volcanic activity.

The developers of this theory had in mind the conjecture made in 1912 by Alfred Wegener (Marburg), who argued from the distribution of species, the shape of continents, and the presence of coastal mountain ranges that the Earth's entire landmass once fitted together. 'Pangaea' subsequently broke apart and the continents drifted to where we find them. In the modern version of Wegener's theory, it is not the continents, but 'plates' on which they and the oceans ride, that migrate. This theory, 'plate tectonics', received timely support from an entirely disparate line of inquiry into the residual magnetism of rocks. Geologists studying the meanderings of Earth's magnetic poles as recorded in ordinary rocks reached the astonishing conclusion that Earth's field has reversed from time to time, the south pole going north and the north south. The most recent of these reversals took place less than a million years ago.

In 1963, Fred Vine and Drummond Matthews (Cambridge) and Lawrence Morley (Canadian Geological Survey) suggested that the sea floor had recorded the magnetic reversals in stripes parallel to the vents from which the molten rock emerges. And this is what the oceanographic research ship *Eltanin*, operated by Columbia's Lamont Geophysical Laboratory, discovered. The travels of the continents, as revealed with the help of sensitive magnetometers, mass spectrometers, argon-potassium radioactive dating, and other instrumentation derived from military technology, take place at the brisk tempo of 4.5 cm per year.

The resources of some international facilities—like CERN and the European Space Agency, high-profile national ones like Russia's space programme, and some industrial research laboratories—have long challenged American supremacy in physics. More recently, Japan, China, and India have added their appreciable weight to the competition. Nonetheless, the US still dominates research. This is not merely because of the amount of money that has poured into its universities and national laboratories from federal and state government, the military, and industry. American dominance will probably continue for a time even as international pressures progressively reduce it. US research universities train foreign graduate students and postdocs as well as domestic ones, and do so efficiently in the approximation to English that is the international language of science.

Chapter 7
The quintessential

For twenty centuries and more, deduction of the place of God and man in the universe was an accepted, even an expected, activity of *physici*. Perhaps the last significant expression of this corporate responsibility occurred in England around the middle of the 19th century in eight volumes commissioned by the Royal Society of London under a bequest from the last Earl of Bridgewater. The volume demonstrating via physics 'the power, wisdom, and goodness of God, as manifested in the creation' was the work of the same William Whewell who invented the term 'physicist'. Many individual physicists since have tried to unite traditional belief with their science, for example, Pascual Jordan, who found room for the freedom of the will in Bohr's interpretation of quantum mechanics.

Bohr applied complementarity in the Greek direction, obtaining insight into the human condition from his understanding of physics. According to him, just as physicists cannot give a complete description of their experiments without invoking formally contradictory concepts, like wave and particle, so philosophers must rely on opposing concepts, like free will and determinism, for an exhaustive description of the world. And just as no realizable experimental arrangement can require a quantum entity to behave simultaneously as a wave and a particle, we cannot experience freedom and determinism together. Freedom applies in prospect, determinism in retrospect, religion not at all. Bohr once intended to write a book against religion to fulfil the moral obligation that his understanding of physics placed on him. Similarly, Einstein's constant challenge to Bohr's 'tranquilizing philosophy or religion' responded to a moral imperative. It was not a reactionary objection to novelty, but a duty to science, to try to maintain the ideal of space-time description. His God who does not play dice represented this ideal: Einstein rejected the concept of a personal deity and an afterlife as firmly as Bohr did.

The modern evolution of the physicists' world picture continues the ancient programme of locating human beings in a directionless universe. We learn that the explosion that created it had to be nicely balanced to produce a state suitable for life. If too forceful, the primitive ingredients would have been scattered too far to form stars and galaxies; if too weak, the universe would have lasted only a second or two before sinking back into nothing. Apparently the initial power exercised, and the strengths of the forces subsequently at work, supplied the time needed (some fifteen billion years) to develop intelligent life, which now uses its own existence as a criterion in designing accounts of the cosmos. This 'anthropic hypothesis', so named by Brandon Carter (Cambridge) in 1970, tickled the fancies of strong speculative minds, gave comfort to believers who saw the hand of God in the fine-tuning of the initial conditions, and appealed to atheists who interpreted our universe as a statistically improbable case of all possible universes originating in big bangs.

Another inference, implying a Bungling Demiurge rather than an Intelligent Creator, comes from the collection of investigations known as SETI (search for extraterrestrial intelligence). The US Congress voted a substantial sum to look for evidence of civilizations capable of producing electromagnetic messages. Not a peep has been heard despite the plausible assumption that among the billions and billions of possible habitats, some must have grown other beings during the last fifteen billion years capable of inventing television. Is the universe not isotropic? Why then the silence? Perhaps the advanced civilizations we might imagine have had the foresight and skill to cordon off our corner of the universe. Or, perhaps more likely, electromagnetic civilizations do not survive long enough to give us a chance to tune into ET TV.

The universe as interpreted by modern physical science is horrifying as well as inspiring whether we are alone in it or not. Earth, with its crashing plates, earthquakes, eruptions, hurricanes, and tsunamis is constantly exposed to bombardment from large meteors capable of wiping out most advanced life forms, as happened here only sixty-six million years ago. If a meteor or comet does not get us, we are sure to be destroyed by our own Sun. The heavens, which Aristotle pictured as a domain of eternal serenity, turn out to be filled with suicidal stars, black holes, super-energetic quasars, forbidding temperatures, and deadly radiations. Physicists believe that the universe originated in a catastrophic explosion of unimaginable violence owing to a chance fluctuation of nothing at all that scattered stars in their billions, and dark matter in its veil, through an indefinitely expanding or pulsating space. What, to repeat the quintessential question, is the place of humankind in this awesome world?

Steven Weinberg, who won a Nobel Prize for his part in advancing a TOE, gives the obvious adult answer: 'The more we know about the universe, the more it is evident that it is pointless and meaningless'. That is no reason not to try to comprehend it, however. Weinberg again: 'The effort to understand the universe is one of the very few things that lifts human life above the level of farce, and gives it some of the grace of tragedy'. In this telling, the purpose of the gigantic, finely tuned experiment is to give life the time and environment to evolve into physicists capable of developing theories of everything and inventing electromagnetic civilizations incapable of survival.

While awaiting the end, physicists pursuing modern versions of Greek speculations have exciting problems to solve. David Gross (Santa Barbara) has made a list of them. Here is the first: 'can we actually say what happened at the beginning of the universe?' Hardly. We cannot even say what is happening now, since we are largely ignorant of the nature of the dark matter supposed to make up most of the mass of the universe. Perhaps CERN's upgraded Large Hadron Collider will make gluinos and neutralinos, the superparticles that supersymmetry offers as candidate constituents of the dark filling of the heavens. Physicists who remember Aristotle call this filling 'quintessence'. Gravity still eludes efforts to unite it with the three other fundamental forces. Nonetheless the theory may be on the right track. In 2016, the US's Advanced LIGO (Laser Interferometer Gravitational-Wave Observatory) detected the gravity waves that general relativity predicted as a result of the mating of black holes.

Why (Gross continued) do the fundamental parameters of the physicists' world picture have the values we find? No Pythagoras has arisen to explain why the top quark (whose existence was established at Fermilab after the inspection of 500 trillion collisions) weighs 100,000 times as much as the up quark, or the proton almost 2,000 times as much as the electron, or why nature has been so generous as to provide physicists with three families each of quarks and leptons (electrons, light mesons, neutrinos). Why is the proton stable? Why does the universe contain so little antimatter? Are the universal constants really constant? Is the concept of space-time fundamental? Does the latest fad—string theory—have the answers? Perhaps. But, alas, Gross writes, 'we really do not understand what string theory is'.

Is there time to learn?

De-deifying and de-anthropomorphizing nature produced physics, and physics now claims control of large swathes of territory previously ruled by capricious deities and organized religion.

If the consequent picture of nature alarms humankind, this might reflect the healthy recognition that the world was not designed for us. The discoveries that no power made the universe for our pleasure, exploitation, or fright, and that there is no divine plot to history, can be liberating for people with the courage to accept them. Physics has given civilization a sombre, disturbing, and challenging world picture, many fertile and some terrifying inventions, and notice of responsibility for the outcome of the human story. If humankind accepts the responsibility and the concomitant loss of providential deities and sacred dicta, the human species might beat the odds against the survival of an electromagnetic civilization, preserve the Earth, and, in the fullness of time, arrive at several satisfactory theories of everything.

References

Introduction: the Greek way

Glashow, Sheldon, and Leon Lederman. 'The SSC: A Machine for the Nineties'. *Physics Today* 38/3 (1985), 28–37.

Chapter 1: Invention in antiquity

'the source or cause': Aristotle, *Phys.*, 192b22. In *The Works of Aristotle Translated into English*. Ed. W. D. Ross. 12 vols. Oxford: Oxford University Press, 1910–52.

'Of all who have discussed': Aristotle, *Metaphys.*, 988a. In *Works*, ed. Ross.

'rule [that] applies': Aristotle, *Phys.*, 198a18. In *Works*, ed. Ross.

'Same stuff': Aristotle, *Meteor.*, 370a25–6. In *Works*, ed. Ross.

'It is fitting': Francis M. Cornford, *Plato's Cosmology: The Timaeus of Plato*. Indianapolis: Bobbs-Merrill, n.d. (The Library of Liberal Arts, 101.), quoting *Timaeus*, 29D.

'The great world's origin': Ovid, *Metamorphoses*. Tr. A. D. Melville. Oxford: Oxford University Press, 2008, p. 366.

'Many philosophical lineages': Seneca, *Natural Questions*. Tr. Harry M. Hine. Chicago: Chicago University Press, 2010, p. 135.

'[T]he rapidity of its motion' and 'the position of earth': Plutarch, *De facie quae in orbe lunae apparet*. Ed. Harold Cherniss. In Plutarch *Moralia*, XII. Cambridge, MA: Harvard University Press, 1995, pp. 59, 75.

'mere accessories' and 'every act': quoted in G. E. R. Lloyd, *Ancient Culture and Society*. 2 vols. Cambridge: Chatto & Windus, 1970–3, 2, pp. 92, 94.

'corruptors and destroyers': quoted in Lloyd, *Ancient Culture and Society*, 2, p. 94.
'courteous, just, and honest,' 'principles of physics,' and 'dig down': Vitruvius, *The Ten Books on Architecture*. Tr. Morris Hicky Morgan. Cambridge, MA: Harvard University Press, 1914, p. 21.
'strange and wonderful': pseudo-Aristotle, *Mech.*, 855b25–30. In Aristotle, *Works*, ed. Ross.
'There was never any': Pliny the Elder, *The Historie of the World: Commonly Called the Naturall Historie*. Tr. Philemon Holland. 2 vols. London: A. Islip, 1635, 2, p. 586.
'disengaged from': Macrobius, *Commentary on the Dream of Scipio*. Tr. W. H. Stahl. New York: Columbia University Press, 1952, p. 171.
'circling by their own surging': W. H. Stahl, Richard Johnson, and E. L. Burge, *Martianus Capella and the Seven Liberal Arts*. 2 vols. New York: Columbia University Press, 1971–7, *1*, pp. 176–7, *2*, pp. 317–19.

Chapter 2: Selection in Islam

Muhammad Ibn Tufayl. *The History of Hai Eb'n Yockdan, an Indian Prince: or, the Self-taught Philosopher*. London: R. Chiswell, 1686.
'I evermore': Omar Khayyam, *Rubáiát*, verse 27. Tr. Edward Fitzgerald. Ed. Daniel Karlin. Oxford: Oxford University Press, 2009, p. 29.
'observation of the stars': Régis Morelon, 'Panorama général de l'histoire de l'astronomie arabe'. In Rashdi Rashed, ed., *Histoire des sciences arabes*. 3 vols. Paris: Seuil, 1997, *1*, p. 30.
'like snow upon the desert': Khayyam, *Rubáiát*, verse 14. Ed. Karlin, p. 23.
The Qur'an. Tr. M. A. S. Abdel Haleem. Oxford: Oxford University Press, 2008.
'their instruments were more precise': Hakim Mohammed Said and Ansar Zahid Khan, *Al Bīrunī: His Times, Life, and Works*. Karachi: Hamdard Foundation, 1981, p. 70.
'the most important document': Ahmad Y. Al-Hassan and Donald R. Hill, *Islamic Technology: An Illustrated History*. Paris: UNESCO, 1986, pp. 58–9.

Chapter 3: Domestication in Europe

Paul Tannery. 'Une correspondance d'écolatres du xi. siècle'. In Tannery. *Mémoires scientifiques*. 17 vols. Toulose: E. Privat, 1912–50, *5*, pp. 103–11.

'the ocean tide': quoted from Pierre Duhem, *Le système du monde: histoire des doctrines cosmologiques de Platon à Copernic*. 10 vols. Paris: Hermann, 1954–9, *3*, pp. 16–20; 'in 800 years': C. C. Gillispie et al., eds., *Dictionary of Scientific Biography*. 18 vols. New York: Scribners, 1970–81, *1*, p. 565.

'the Arabic pomposity': Dorothea Metlitzki, *The Matter of Araby in Medieval England*. New Haven: Yale University Press, 1977, p. 5, quoting the Bishop of Cordoba.

'our civilization': Metlitzki, *The Matter of Araby in Medieval England*, p. 11, quoting Robert of Morley.

'knowing nothing': 'Epistola Pharamellae,' in Roger de Hoveden, *Chronica*. Ed. William Stubbs. 4 vols. London: Longman et al., 1868–71, *2*, pp. 296–8.

'wholly unknown': quoted from Bacon's *Opus maius* by A. C. Crombie and John North, in Gillispie et al., *Dictionary of Scientific Biography*, *1*, p. 380.

'It shall come to pass': *Genesis*, 8: 13, 14, 16.

'no less than' and 'angelic art': quoted from Noel Swerdlow, 'Science and Humanism in the Renaissance: Regiomontanus' Oration on the Dignity and Utility of the Mathematical Sciences'. In Paul Harwich, ed., *World Changes: Thomas Kuhn and the Nature of Science*. Cambridge, MA: MIT Press, 1993, pp. 133, 165.

'man is a *magnum miraculum*': quoted from Frances A. Yates, *Giordano Bruno and the Hermetic Tradition*. Chicago: Chicago University Press, 1964, p. 35.

'intoxicated, crazy': William Gilbert, *De magnete*. Tr. P. M. Mottelay. New York: Dover, 1958, pp. xlviii, xlix.

Chapter 4: Second creation

'New philosophy calls all in doubt': Donne, quoted from Marjorie Nicholson, *Science and Imagination*. Ithaca: Cornell University Press, 1956, p. 53.

Francis Bacon. *The Major Works*. Ed. Brian Vickers. Oxford: Oxford University Press, 2002.

René Descartes. *Discourse on Method, Optics, Geometry, and Meteorology*. Tr. Paul J. Olscamp. Indianapolis: Bobbs-Merrill, 1965. (Library of Liberal Arts, 211.)

'oddest...detection': Newton to Oldenburg, 18 January 1671–2, in Isaac Newton, *The Correspondence of Isaac Newton*. Ed. H. W. Turnbull et al. 7 vols. Cambridge: Cambridge University Press, 1959–77, *1*, pp. 82–3.

'from the counsel,' 'I frame no hypothesis,' and '[T]o us it is enough': Isaac Newton, *Mathematical Principles of Natural Philosophy*. Tr. Andrew Motte (1729), revised Florian Cajori. Berkeley: University of California Press, 1934, pp. 544–7.

'in every special doctrine' and 'Chemistry can become nothing': Immanuel Kant, *Metaphysical Foundations of Modern Science*. Tr. James W. Ellington. Indianapolis: Bobbs-Merrill, 1970, pp. 6, 7.

'thermometer of public prosperity': Karin Johannison, 'Society in Numbers: The Debate Over Quantification in 18th-century Political Economy'. In Tore Frängsmyr, J. L. Heilbron, and Robin E. Rider, eds., *The Quantifying Spirit in the 18th Century*. Berkeley: University of California Press, p. 358, quoting Anon., *Recherches...sur la population de la France* (1778).

'Experiment, research': J. L. Heilbron, 'The measure of Enlightenment'. In Frängsmyr, Heilbron, and Rider, eds., *The Quantifying Spirit in the 18th Century*, pp. 10–11, quoting Lavoisier.

Chapter 5: Classical physics and its cure

Henri Poincaré. 'Relations entre la physique expérimentale et la physique mathématique'. In Congrès international de physique réuni à Paris en 1900. *Rapports et travaux*. 4 vols. Paris: Gautier-Villars, 1900–1, 1, pp. 1–29.

William Thomson and Baron Kelvin. 'Nineteenth Century Clouds over the Dynamical Theory of Heat and Light' (1900). In Thomson, *Baltimore Lectures on Molecular Dynamics and the Wave Theory of Light*. London: C.J. Clay, 1904, pp. 486–527.

'who, during the preceding year': Nobel Foundation, *Directory*. Stockholm: Nobel Foundation, annual, p. 5.

'Science has reached a point': 'Denkschrift...an den Kaiser,' in Kaiser-Wilhelm-Gesellschaft, *50 Jahre Kaiser-Wilhelm-Gesellschaft und Max-Planck-Gesellschaft: Beiträge und Dokumente*. Göttingen: Max-Planck-Gesellschaft, 1961, p. 90.

Chapter 6: From old world to new

'the elements originated': quoted in Helge Kragh, *Cosmology and Controversy: The Historical Development of Two Theories of the Universe*. Princeton: Princeton University Press, 1996, p. 105.

Chapter 7: The quintessential

'tranquilizing philosophy': Einstein to Schrödinger, May 31, 1928, in Albert Einstein, et al., *Letters on Wave Mechanics*. New York: Philosophical Library, 1967, p. 31.

'The more we know' and 'The effort to understand': Steven Weinberg, *The First Three Minutes: A Modern View of the Origin of the Universe*. 2nd edn. New York: Basic Books, 1993, pp. 154–5.

'can we actually say' and 'we really do not understand': David Gross, 'The Major Unknowns in Particle Physics and Cosmology'. In R. Y. Chiao et al., eds., *Visions of Discovery: New Light on Physics, Cosmology, and Consciousness*. Cambridge: Cambridge University Press, 2011, pp. 153, 165.

Further reading

General

Bearman, Peri J. et al., eds. *The Encyclopedia of Islam*. Leyden: Brill, 2000–. Multivolume electronic resource.

Boyer, Carl B. *The Rainbow: From Myth to Mathematics*. Basingstoke: Macmillan, 1987.

Buchwald, Jed, and Robert Fox, eds. *The Oxford Handbook of the History of Physics*. Oxford: Oxford University Press, 2013.

Bush, Steven. *A History of Modern Planetary Physics*. 3 vols. Cambridge: Cambridge University Press, 1996.

Darrigol, Olivier. *A History of Optics from Greek Antiquity to the 19th Century*. Oxford: Oxford University Press, 2012.

Dijksterhuis, E. J. *The Mechanization of the World Picture*. Oxford: Oxford University Press, 1961.

Frisinger, H. Howard. *The History of Meteorology to 1800*. New York: Science History Publications, 1977.

Gillispie, C. C. et al., eds. *Dictionary of Scientific Biography*. 18 vols. New York: Scribners, 1970–81; continued by Noretta Koertge, ed., *New DSB*, 8 vols, 2008.

Heilbron, J. L., ed. *The Oxford Companion to the History of Modern Science*. Oxford: Oxford University Press, 2003.

Kragh, Helge. *Conceptions of Cosmos: From Myths to the Accelerating Universe*. New York: Oxford University Press, 2007.

Kragh, Helge. *Higher Speculations: Grand Theories and Failed Revolutions in Physics and Cosmology*. Oxford: Oxford University Press, 2011.

Lindberg, D. C., and R. L. Numbers et al., eds. *The Cambridge History of Science*. 6 vols (of 8). Cambridge: Cambridge University Press, 2003–.

North, John. *Cosmos: An Illustrated History of Astronomy and Cosmology*. Chicago: University of Chicago Press, 2008.

Petruccioli, Sandro et al., eds. *Storia della scienza*. 10 vols. Rome: Istituto della Enciclopedia Italiana, 2001–4.

Rossi, Paolo et al., eds. *Storia della scienza moderna e contemporanea*. 6 vols. Milan: TEA, 2000.

The (US) History of Science Society publishes an indispensable guide to literature in the history of science, now available online at <IsisCB.org>.

Chapter 1: Invention in antiquity

Cohen, Morris R., and I. E. Drabkin, eds. *A Source Book in Greek Science*. Cambridge, MA: Harvard University Press, 1958.

Lucretius. *On the Nature of Things*. Tr. Ronald Melville. Oxford: Oxford University Press, 2008.

Ptolemy. *Tetrabiblos*. Tr. F. E. Robbins. Cambridge, MA: Harvard University Press, 1964.

Stahl, William H. *Roman Science: Origins, Development and Influence to the Later Middle Ages*. Madison: University of Wisconsin Press, 1962.

Taub, Liba. *Ancient Meteorology*. London: Routledge, 2003.

Chapter 2: Selection in Islam

Bloom, Jonathan M. *Paper before Print: The History and Impact of Paper in the Islamic World*. New Haven: Yale University Press, 2001.

Ihsanoglu, Ekmeleddin. *Science, Technology and Learning in the Ottoman Empire*. Aldershot: Ashgate, 2004.

Makdisi, George. *The Rise of Colleges: Institutions of Learning in Islam and the West*. Edinburgh: Edinburgh University Press, 1981.

Makdisi, George. *The Rise of Humanism in Classical Islam and the Christian West: With Special Reference to Scholasticism*. Edinburgh: Edinburgh University Press 1990.

Masood, Ehsan. *Science and Islam: A History*. London: Icon, 2009.

Nasr, Sayyid Hussein. *Islamic Science: An Illustrated Survey*. World of Islam, 1976.

Peters, F. E. *Aristotle and the Arabs*. New York: New York University Press, 1968.
Rashed, Rashdi, ed. *Histoire des sciences arabes*. 3 vols. Paris: Seuil, 1997.
Saliba, George. *Islamic Science and the Making of the European Renaissance*. Cambridge, MA: MIT Press, 2007.
Sayili, Aydin. *The Observatory in Islam*. New York: Arno, 1981.
Turner, Howard R. *Science in Medieval Islam: An Illustrated Introduction*. Austin: University of Texas Press, 1997.

Chapter 3: Domestication in Europe

Clagett, Marshall. *The Science of Mechanics in the Middle Ages*. Madison: University of Wisconsin Press, 1959.
Copernicus, Nicolas. *On the Revolutions*. Tr. Edward Rosen. Baltimore: Johns Hopkins University Press, 1978.
Dante Alighieri. *The Divine Comedy*. Ed. Charles Singleton. 6 vols. Princeton: Princeton University Press, 1970–5.
Maier, Annaliese. *Studien zur Naturphilosophie der Spätscholastik*. 5 vols. Rome: Storia e letteratura, 1949–58.
Yates, Frances A. *Giordano Bruno and the Hermetic Tradition*. Chicago: University of Chicago Press, 1964.

Chapter 4: Second creation

Frängsmyr, Tore, J. L. Heilbron, and Robin E. Rider, eds. *The Quantifying Spirit in the 18th Century*. Berkeley: University of California Press, 1990.
Friedman, Michael. *Kant and the Exact Sciences*. Cambridge, MA: Harvard University Press, 1992.
Heilbron, J. L. *Electricity in the 17th and 18th Centuries: A Study in Early Modern Science*. Berkeley: University of California Press, 1979.
Taton, René, ed. *Enseignement et diffusion des sciences en France au xviiie siècle*. Paris: Hermann, 1964.

Chapter 5: Classical physics and its cure

Buchwald, Jed. *From Maxwell to Microphysics: Aspects of Electromagnetic Theory in the Last Quarter of the 19th Century*. Chicago: University of Chicago Press, 1985.

Buchwald, Jed. *The Rise of the Wave Theory of Light: Optical Theory and Experiment in the Early 19th Century*. Chicago: University of Chicago Press, 1989.
Congrès international de physique réuni à Paris en 1900. *Rapports et travaux*. 4 vols. Paris: Gautier-Villars, 1900–1.
Darrigol, Olivier. *Electrodynamics from Ampère to Einstein*. Oxford: Oxford University Press, 2000.
Forman, Paul, J. L. Heilbron, and Spencer Weart. *Physics circa 1900*. Princeton: Princeton University Press, 1975. (Historical Studies in the Physical Sciences, vol. 5.)

Chapter 6: From old world to new

Dahl, Per F. *Superconductivity: Its Historical Roots and Development from Mercury to the Ceramic Oxides*. New York: American Institute of Physics, 1992.
Fleming, James Rodger, ed. *Historical Essays on Meteorology, 1919–1995*. Boston: American Meteorological Society, 1996.
Forman, Paul. *National Military Establishments and the Advancement of Science and Technology: Studies in 20th Century History*. Ed. J. M. Sanchez Ron. Dordrecht: Kluwer, 1996.
Forman, Paul. *Weimar Culture and Quantum Mechanics: Selected Papers*. Ed. Cathryn Carson et al. London: Imperial College Press, 2011.
Glen, William. *The Road to Jaramillo: Critical Years of the Revolution in Earth Science*. Stanford: Stanford University Press, 1982.
Heilbron, J. L., and Robert W. Seidel. *Lawrence and his Laboratory*. Berkeley: University of California Press, 1989.
Krige, John. *American Hegemony and the Postwar Reconstruction of Science in Europe*. Cambridge: MIT Press, 2006.
Krige, John, and Dominique Pestre, eds. *Science in the Twentieth Century*. Amsterdam: Harwood, 1997.
Mendelssohn, Kurt. *The Quest for Absolute Zero*. London: Taylor & Francis, 1977.

Index

A

B

C

D

E

F

G

H

I

J

K

L

M

N

O

P

Q

R

S

T

U

V

W

Y

Z

NOTHING
A Very Short Introduction
Frank Close

What is 'nothing'? What remains when you take all the matter away? Can empty space - a void - exist? This *Very Short Introduction* explores the science and history of the elusive void: from Aristotle's theories to black holes and quantum particles, and why the latest discoveries about the vacuum tell us extraordinary things about the cosmos. Frank Close tells the story of how scientists have explored the elusive void, and the rich discoveries that they have made there. He takes the reader on a lively and accessible history through ancient ideas and cultural superstitions to the frontiers of current research.

> 'An accessible and entertaining read for layperson and scientist alike.'
>
> **Physics World**

www.oup.com/vsi